D0721650

Exploring Organic Environments in the Solar System

Task Group on Organic Environments in the Solar System

Space Studies Board
Division on Engineering and Physical Sciences
and
Board on Chemical Sciences and Technology
Division on Earth and Life Studies

NATIONAL RESEARCH COUNCIL
OF THE NATIONAL ACADEMIES

THE NATIONAL ACADEMIES PRESS
Washington, D.C.
www.nap.edu

THE NATIONAL ACADEMIES PRESS • **500 Fifth Street, N.W.** • **Washington, DC 20001**

NOTICE: The project that is the subject of this report was approved by the Governing Board of the National Research Council, whose members are drawn from the councils of the National Academy of Sciences, the National Academy of Engineering, and the Institute of Medicine. The members of the task group responsible for the report were chosen for their special competences and with regard for appropriate balance.

This study was supported by Contracts NASW-96013 and NASW-01001 between the National Academy of Sciences and the National Aeronautics and Space Administration. Any opinions, findings, conclusions, or recommendations expressed in this publication are those of the author(s) and do not necessarily reflect the views of the agency that provided support for the project.

Cover: An artist's impression of the European Space Agency's Huygens probe sitting in an apparent floodplain of icy pebbles soon after its successful landing on the surface of Saturn's giant moon, Titan, on January 14, 2005. Data from Huygens and its companion, the Cassini orbiter, have provided new insights into the degree of chemical complexity that can arise from 4.5 billion years of planetary evolution in an environment rich in organic compounds. Image courtesy of the European Space Agency.

International Standard Book Number-13 978-0-309-10235-3
International Standard Book Number-10 0-309-10235-9

Copies of this report are available free of charge from:

Space Studies Board
National Research Council
500 Fifth Street, N.W.
Washington, DC 20001

Additional copies of this report may be purchased from the National Academies Press, 500 Fifth Street, N.W., Lockbox 285, Washington, DC 20055; (800) 624-6242 or (202) 334-3313 (in the Washington metropolitan area); Internet, http://www.nap.edu.

THE NATIONAL ACADEMIES
Advisers to the Nation on Science, Engineering, and Medicine

The **National Academy of Sciences** is a private, nonprofit, self-perpetuating society of distinguished scholars engaged in scientific and engineering research, dedicated to the furtherance of science and technology and to their use for the general welfare. Upon the authority of the charter granted to it by the Congress in 1863, the Academy has a mandate that requires it to advise the federal government on scientific and technical matters. Dr. Ralph J. Cicerone is president of the National Academy of Sciences.

The **National Academy of Engineering** was established in 1964, under the charter of the National Academy of Sciences, as a parallel organization of outstanding engineers. It is autonomous in its administration and in the selection of its members, sharing with the National Academy of Sciences the responsibility for advising the federal government. The National Academy of Engineering also sponsors engineering programs aimed at meeting national needs, encourages education and research, and recognizes the superior achievements of engineers. Dr. Wm. A. Wulf is president of the National Academy of Engineering.

The **Institute of Medicine** was established in 1970 by the National Academy of Sciences to secure the services of eminent members of appropriate professions in the examination of policy matters pertaining to the health of the public. The Institute acts under the responsibility given to the National Academy of Sciences by its congressional charter to be an adviser to the federal government and, upon its own initiative, to identify issues of medical care, research, and education. Dr. Harvey V. Fineberg is president of the Institute of Medicine.

The **National Research Council** was organized by the National Academy of Sciences in 1916 to associate the broad community of science and technology with the Academy's purposes of furthering knowledge and advising the federal government. Functioning in accordance with general policies determined by the Academy, the Council has become the principal operating agency of both the National Academy of Sciences and the National Academy of Engineering in providing services to the government, the public, and the scientific and engineering communities. The Council is administered jointly by both Academies and the Institute of Medicine. Dr. Ralph J. Cicerone and Dr. Wm. A. Wulf are chair and vice chair, respectively, of the National Research Council.

www.national-academies.org

OTHER REPORTS OF THE SPACE STUDIES BOARD

NOTE: Listed according to year of approval for release, which in some cases precedes the year of publication.

TASK GROUP ON ORGANIC ENVIRONMENTS IN THE SOLAR SYSTEM

JAMES P. FERRIS, Rensselaer Polytechnic Institute, *Chair*
LUANN BECKER, University of California, Santa Barbara
KRISTIE A. BOERING, University of California, Berkeley
GEORGE D. CODY, Carnegie Institution of Washington
G. BARNEY ELLISON, University of Colorado
JOHN M. HAYES, Woods Hole Oceanographic Institution
ROBERT E. JOHNSON, University of Virginia
WILLIAM KLEMPERER, Harvard University
KAREN J. MEECH, University of Hawaii
KEITH S. NOLL, Space Telescope Science Institute
MARTIN SAUNDERS, Yale University

Staff

DAVID H. SMITH, Study Director
SANDRA J. GRAHAM, Senior Staff Officer, Space Studies Board
ROBERT L. RIEMER, Senior Staff Officer, Space Studies Board
CHRISTOPHER K. MURPHY, Senior Staff Officer, Board on Chemical Sciences and Technology
CATHERINE A. GRUBER, Assistant Editor
RODNEY HOWARD, Senior Project Assistant

Preface

The sources, distributions, and transformations of organic compounds throughout diverse solar system environments are active areas of scientific study. Although numerous methods of organic synthesis and/or alteration have been demonstrated using a wide variety of starting materials, energy sources, and environments, the signatures that would make it possible to discriminate among these processes are not well established. As a result, ambiguity surrounds the relative effectiveness of various organic-synthesis processes in explaining the distribution of organic compounds in the solar system.

To study these issues in more detail, the Task Group on Organic Environments in the Solar System (TGOESS)—an ad hoc group consisting of members drawn primarily from the Committee on Planetary and Lunar Exploration (COMPLEX), the Committee on the Origins and Evolution of Life, and the Board on Chemical Sciences and Technology—was established. In particular, the task group was asked to determine what processes account for the reduced carbon compounds found throughout the solar system and to examine how planetary exploration can advance understanding of this central issue. In addition, the task group was asked to consider a number of closely related questions, including the following:

1. What are the sources of reactants and energy that lead to abiotic synthesis of organic compounds and to their alteration in diverse solar system environments?

2. What are the distribution and history of reduced carbon compounds in the solar system, and which features of that distribution and history, or of the compounds themselves, can be used to discriminate among synthesis and alteration processes?

3. What are the criteria that distinguish abiotic from biotic organic compounds?

4. What aspects of the study of organic compounds in the solar system can be accomplished from ground-based studies (theoretical, laboratory, and astronomical), Earth orbit, and planetary missions (orbiters, landers, and sample return), and which new capabilities might have the greatest impact on each?

Although this project was formally initiated in October 2000, informal presentations in support of it began somewhat earlier and were conducted in the context of COMPLEX's standing oversight of NASA's planetary exploration programs and the definition and development of the charge for this study. These preparatory meetings in March and July 2000 led to the establishment in the summer of 2000 of the Task Group on Organic Environments in the Solar System.

The task group held its first formal meeting at the National Academies' J. Erik Jonsson Woods Hole Center in Woods Hole, Massachusetts, on October 2-4, 2000. The task group's discussions and deliberations continued at meetings held in Tucson, Arizona, and in Washington, D.C., on March 29-31, 2001, and May 2-4, 2001, respectively. An initial draft of the complete report was assembled at a meeting held in Irvine, California, on November 17-21, 2003. A draft of the report was sent to external reviewers in May 2004. The text was extensively revised, updated, and finalized in the first half of 2006 and was approved for release by the National Research Council (NRC) on November 21, 2006.

The work of the task group was made easier thanks to the contributions made by Louis Allamandola (NASA, Ames Research Center), Richard Binzel (Massachusetts Institute of Technology), Geoffrey A. Blake (California Institute of Technology), Sherwood Chang (SETI Institute), John Cronin (Arizona State University), Dale Cruikshank (NASA, Ames Research Center), Pascale Ehrenfreund (Leiden Observatory), George Flynn (State University of New York, Plattsburgh), Randy Gladstone (Southwest Research Institute), William Irvine (University of Massachusetts, Amherst), Paul Lucey (University of Hawaii), Jonathan Lunine (University of Arizona), Scott Messenger (Washington University), Elisabetta Pierazzo (University of Arizona), Wayne Roberge (Rensselaer Polytechnic Institute), and Lucy Ziurys (University of Arizona).

The task group also expresses its appreciation to Everett Shock for persuading COMPLEX of the importance of the issues discussed in this report, for taking the leading role in developing and drafting the charge for this study, and for paving the way for the work of the task group.

This report has been reviewed by individuals chosen for their diverse perspectives and technical expertise, in accordance with procedures approved by the NRC's Report Review Committee. The purpose of this independent review is to provide candid and critical comments that will assist the authors and the NRC in making the published report as sound as possible and to ensure that the report meets institutional standards for objectivity, evidence, and responsiveness to the study charge. The contents of the review comments and draft manuscripts remain confidential to protect the integrity of the deliberative process. We wish to thank the following individuals for their review of this report: Charles H. DePuy, University of Colorado, Boulder; Thomas M. Donahue, University of Michigan, Ann Arbor; Eric Herbst, Ohio State University; Donald M. Hunten, University of Arizona; Timothy J. McCoy, Smithsonian Institution; Harry Y. McSween, Jr., University of Tennessee; and Anna-Louise Reysenbach, Portland State University.

Although the reviewers listed above have provided many constructive comments and suggestions, they were not asked to endorse the conclusions or recommendations, nor did they see the final draft of the report before its release. The review of this report was overseen by Bernd R.T. Simoneit, Oregon State University. Appointed by the NRC, he was responsible for making certain that an independent examination of this report was carried out in accordance with institutional procedures and that all review comments were carefully considered. Responsibility for the final content of this report rests entirely with the authoring task group and the institution.

Contents

Executive Summary

The sources, distributions, and transformations of organic compounds throughout the solar system are being studied actively. The results can provide information about the evolution of the solar system and about possibilities for life elsewhere in the universe. All life on Earth is based on the complex interplay of diverse carbon compounds. In short, the chemistry of carbon is the chemistry of life. But carbon is extremely versatile. Its compounds can be synthesized in many different ways and from a wide variety of starting materials, the vast majority of which have nothing to do with biology. The chemical reactions involving carbon can be driven by many different sources of energy and can occur in diverse environments, many of which are inimical to life as we understand it. Many carbon compounds are extremely hardy, and their preservation in the geological record can tell researchers much about processes and environmental conditions in the distant past. Similarly, other carbon compounds are extremely fragile. The presence of organic compounds in various astronomical environments can tell researchers much about the conditions that prevail today.

The discovery of a single drop of oily residue on Mars, for example, would be enormously informative even if the residue were irrefutably abiotic in origin. Some might argue that such a discovery would be like finding an encyclopedia from a Mars library, which would tell linguists so much about the inhabitants even if they could not translate it. To recover the information carried by extraterrestrial carbon compounds, researchers must improve their ability to recognize the signals that point to specific syntheses and conditions.

PURPOSE AND APPROACH OF THIS REPORT

The purpose of this report is to tell the story of carbon: to follow carbon through a variety of terrestrial and extraterrestrial environments, to track its changes as it is subjected to a variety of physical and chemical processes, and to attempt to convey what the study of carbon and its compounds tells us about the origin and evolution of the solar system. In particular, the Task Group on Organic Environments in the Solar System surveys what is known about the sources of reduced carbon compounds throughout the solar system and examines how planetary exploration can improve our understanding. It is not the purpose of this report to recommend expensive new research activities and propose costly new initiatives. Rather, the task group's goal is to place a variety of disparate activities in a unified context. As part of this process, the task group considers a number of closely related questions, including the following:

1. What are the sources of reactants and energy that lead to abiotic synthesis of organic compounds and to their alteration in diverse solar system environments?

2. What are the distribution and history of reduced carbon compounds in the solar system, and which features of that distribution and history, or of the compounds themselves, can be used to discriminate among synthesis and alteration processes?

3. What are the criteria that distinguish abiotic from biotic organic compounds?

4. What aspects of the study of organic compounds in the solar system can be accomplished from ground-based studies (theoretical, laboratory, and astronomical), Earth orbit, and planetary missions (orbiters, landers, and sample return), and which new capabilities might have the greatest impact on each?

The task group found it most convenient and logical to address the third question first. The reason for this approach is simple. The principal features that distinguish biotic and abiotic carbon compounds are closely related to the physical and chemical characteristics of organic compounds. Thus, these distinguishing criteria are elaborated in the context of a general introduction to organic chemistry in Chapter 1. With regard to the indicators that might differentiate between a biotic and an abiotic origin for particular organic compounds, the task group found that the most compelling indicators of an abiotic origin include the following:

- The presence of a smooth distribution of organic compounds in a sample, e.g., a balance of even versus odd numbers of carbon atoms in alkanes;
- The presence of all possible structures, patterns, isomers, and stereoisomers in a subset of compounds such as amino acids;
- A balance of observed entantiomers; and
- The lack of depletions or enrichments of certain isotopes with respect to the isotopic ratio normally expected.

Likewise, the converse of the above items is an indicator of possible biotic synthesis. Thus, for example, an imbalance of even versus odd numbers of carbon atoms in, for example, alkanes or the presence of only a small subset of all possible structures, patterns, isomers, and stereoisomers is an indicator of possible biotic origin. However, some abiotic processes can mimic biotic ones and vice versa, and inferences will necessarily be based on several indicators and will of course be probabilistic.

The answers to the first two questions—sources of reactants and energy that lead to abiotic synthesis and the distribution of organic compounds in the solar system—depend strongly on what part of the solar system is being considered. This report therefore deals separately with the various solar system environments—which range from the surfaces of cold, dark asteroids in remote, eccentric orbits to the hot, turbulent atmospheres of the giant gas planets. It considers what is known about the origins and histories of the organic materials in each setting. This discussion is contained in Chapters 2 through 6 of this report.

The fourth question, research opportunities, is addressed in each of those chapters as well. In addition, Chapter 7 outlines two general strategies recommended by the task group as integral to a planned approach to searching for and understanding organic material in the solar system.

RECOMMENDED RESEARCH

In selecting the best research opportunities for enhancing understanding of organic material in the solar system, the task group considered the following factors:

1. The likelihood that significant organic material would be found;
2. The feasibility of the investigation; and
3. The likely impact or significance of the results.

The recommendations and a brief rationale are given below. A detailed discussion is presented in Chapters 2 through 6 of this report.

Overall Approach to Research

Two recommendations are better characterized as general strategies rather than specific opportunities:

Recommendation: Strategy 1—Every opportunity should be seized to increase the breadth and detail in inventories of organic material in the solar system. As results accumulate, each succeeding investigation should be structured to provide information that will allow improved comparisons between environments. Analyses should determine abundance ratios for the following:

- Compound classes (e.g., aliphatic, aromatic, acetylenic);
- Individual compounds (e.g., methane/ethane);
- Elements in organic material (e.g., C/H/N/O/S); and
- The isotopes of elements such as C, H, N, and O.

Investigators should strive to interpret these results in terms of precursor-product relationships.

These objectives are broadly applicable and represent systematic steps toward addressing questions of biogenicity, lines of inheritance of organic material, and mechanisms of synthesis. With limited funds, returns from investigations like those proposed below (in "Selected Opportunities for Research") will move NASA more smoothly toward ultimate success. For example, the task group proposes that newer, more sensitive, and specific analytical methods be used for the analysis and reanalysis of carbonaceous chondrites. As these studies proceed and the results from flight experiments are obtained, it will become apparent which of these new techniques should be adapted to flight experiments. Moreover, the ground-based investigations of chondrites will pave the way for better analyses of returned samples, whenever they become available.

Recommendation: Strategy 2—Organic-carbon-related flight objectives should be coordinated across missions and structured to provide a stepwise accumulation of basic results. Some of the objectives that should be included in such missions are as follows:

- Quantitation of the amount of organic carbon present to ± 30 percent precision and accuracy over a range of 0.1 parts per million to 1 percent;
- Repetitive analyses of diverse samples at each landing site;
- Comparability so that relatable data are obtained from a wide range of sites; and
- Elemental and isotopic analyses so that the composition (H/C, N/C, O/C, and S/C) is obtained together with the isotope ratios of all the carbon-bearing phases.

These recommended approaches to research will allow scientists to build an overview of the distribution of organic carbon in the solar system; provide information about heterogeneity at each location studied; and support preliminary estimates of relationships, if any, between organic materials at diverse sites.

Selected Opportunities for Research

The selected research opportunities were divided by the task group into three general categories based on the cost of the research and the time frame in which it could be undertaken. The recommended research is given by category below.

Near-Term Opportunities

The first category of research—near-term opportunities—includes ground-based studies that can be carried out in the very near term and for a minimal cost relative to the other recommended research activities.

Chondritic and Mars Meteorites. Carbonaceous meteorites are an important source of abiotic, extraterrestrial carbon that is delivered to Earth at no cost. Together with the unequilibrated ordinary chondrites, a few martian meteorites, and fragments of crust from the earliest Earth, they represent immediately available samples of great relevance to studies of organic material in the solar system. New analyses of carbonaceous chondrites would benefit from modern analytical methods (e.g., compound-specific isotopic analysis) that allow the separation of signals from terrestrial contamination and indigenous extraterrestrial organic matter, thus overcoming a problem that severely hindered analyses throughout the 1960s and 1970s. A more sensitive and detailed analysis of carbonaceous chondrites is a cost-effective step that would be of great value in enhancing understanding of the formation of these organic materials and, therefore, yielding new information about organic-chemical processes in the early solar system. The results would provide reference points for comparison with the organics in samples returned by missions to other bodies in the solar system. Analyses should examine the following:

- The location and relative abundances of the organic molecules within the mineral matrices and on mineral surfaces;
- The structural composition of all organic phases including, to the greatest extent possible, any macromolecular material;
- The isotopic compositions of all molecules and other definable subfractions; and
- The nature of contaminants and the mechanisms by which samples can become contaminated, both before and after collection.

Recommendation: Plans should be developed for the establishment of an informal, community-based forum—modeled on the highly successful Mars Exploration Program Analysis Group (MEPAG)—charged to coordinate plans and develop priorities for the intensive investigation of the composition of organic materials in carbonaceous chondrites, SNC meteorites, and ordinary chondrites containing volatiles (including rare gases) that suggest relationships to the carbonaceous chondrites. The existing Curation and Analysis Planning Team for Extraterrestrial Materials (CAPTEM) may provide the seed from which such a community-based forum can be nurtured.

To provide comparability and to bring the best techniques to bear on each object, samples should be shared extensively between laboratories.

Martian Regolith Simulation. It has been proposed that any organic matter in the martian regolith will have been modified via reaction with strong oxidants present in the soil. Carefully designed laboratory experiments will allow an assessment of this problem and will point to the most effective strategies for direct analysis of organic materials by future Mars landers such as the Mars Science Laboratory. Regolith simulations may help address issues related to, for example, optimal minimum drilling depths for future Mars lander missions.

Recommendation: Laboratory models of Mars soil chemistry should be used to study plausible mechanisms for the oxidative alteration of organic materials in the martian regolith and to evaluate their integrated effects. Materials studied should include likely exogenous products (organic compounds like those found in meteorites) as well as conceivable martian prebiotic and biotic products.

Increasing the Supply of Meteorites Available for Study. The ready availability of and access to meteorites for laboratory studies, particularly the rare carbonaceous chondrites, is a key facet of the exploration of organic environments in the solar system. The preferred means for acquiring samples—collecting them in the field—has led to major searches in those places where meteorites are most likely to be spotted, the hot and cold deserts of the world. Both locations have their advantages and disadvantages; a detailed cost-benefit analysis of all of the relevant factors is beyond the scope of this report. There is, however, another approach to increasing the supply of meteorites: the selective purchase or exchange of important samples. Indeed, the task group suggests that the greatest near-term scientific impact from a given expenditure of funds will result not from the enhancement of meteorite collecting programs but rather from the acquisition by purchase or exchange of a significant piece of the Tagish Lake meteorite.

Recommendation: The scientific significance of the Tagish Lake meteorite is such that NASA, the National Science Foundation, the Smithsonian Institution, and other relevant organizations and agencies in the United States and their counterparts in Canada should examine the means by which a significant portion of this fall can be acquired, by purchase, exchange, or some other mechanism, so that samples can be made more widely available for study by the scientific community.

Laboratory Studies to Support Observations of Primitive Bodies. Laboratory studies are a prerequisite for all observational studies and are an essential precursor to the design of inherently expensive spacecraft instrumentation. At present, the relevant optical constants have been measured for only a few of the organic and inorganic compounds that are likely to be present in primitive bodies of interest. Without a suite of materials with known constants to incorporate in the spectral models, the identification of many of the observed spectral features remains challenging. With modest support for laboratory work of this kind, great progress could be made in understanding the organic component in these bodies.

Recommendation: The physical, chemical, and spectroscopic properties of ices of potential hydrocarbon species should be studied to facilitate the detection of organic materials.

Support for Telescope Studies of Organic Materials in the Solar System. Access to a small number of unique, publicly available, ground-based infrared astronomical facilities has enhanced and will continue to advance understanding of the organic constituents of various solar system bodies through direct observations and through observations conducted in support of spacecraft missions. Activities that would significantly enhance ground-based observations of organic materials in the solar system include increasing NASA's share of the observing time on the Keck telescope and replacing the NASA Infrared Telescope Facility with a larger instrument capable of making these observations.

Recommendation: The task group reiterates the call made in the 2003 report of the National Research Council's Solar System Exploration Decadal Survey Committee, *New Frontiers in the Solar System,* that NASA's support for planetary observations with ground-based astronomical instruments, such as the Infrared Telescope Facility and the Keck telescopes, be continued and upgraded as appropriate, for as long as they provide significant scientific return and/or mission-critical support services.[1]

Interplanetary Dust and Molecules. Present particle-collection programs utilize aircraft and flights with other primary missions, and schedules are controlled by factors other than the timing of meteor showers. Quantitative yields and the ranges of materials sampled could be greatly improved if flights were timed to utilize these opportunities.

Recommendation: A program specifically designed to collect dust in the stratosphere during meteor showers should be implemented.

Relatively Near-Term Missions Consistent with Previous Decadal Strategy Study

Recommendations in the category of relatively near-term missions are for research that can be implemented or carried out in 5 to 10 years and that is also supported by the findings and recommendations of the 2003 solar system exploration decadal survey report, *New Frontiers in the Solar System.*[2]

Mars. On Earth, the most suitable lithologies for the preservation and accumulation of organic matter are sedimentary rocks that are typically fine-grained and are characterized by well-defined, aqueously derived mineral assemblages. Thus, it may be possible to obtain additional information about the associated organic matter present in these mineral assemblages in a single measurement of the organic and inorganic material present.

Recommendation: Currently planned missions to Mars should seek to identify silicified martian terrains associated with ancient low-temperature hot springs in concert with a high probability of ground ice deposits to locate organic materials formed on Mars. Similarly, the identification of shallow marine and/or lacustrine sediments would provide another terrain well worth exploring in future missions as sites for martian endogenous organosynthesis.

As instrument development continues for future robotic missions to Mars, it is important that such missions be designed so that they are capable of assessing as fully as possible the inventory of organic matter there. Clearly such development should be strongly guided by the information provided by the Mars Exploration Rovers, Spirit and Opportunity.

Although future robotic missions will be equipped with instrumentation to analyze samples, these analyses will never be able to achieve the capabilities of Earth-based laboratories. The discovery by the Mars Exploration Rovers of unambiguous sedimentary outcrops greatly increases the impetus for a martian sample-return mission. Similarly, the discovery of the halogens bromine and chlorine in abundance at the location of the Spirit rover landing site strongly suggests the former presence of surface water. Samples from either location might very well contain organic matter derived from extinct (or perhaps even extant) life. The successes of Spirit and Opportunity further validate the need to implement the 2003 solar system exploration decadal survey's recommendation for a flagship mission to Mars—that is, to begin the developments necessary so that martian samples can be brought back to Earth for study in terrestrial laboratories as early as possible in the next decade.[3]

Far-Term Research Opportunities

The far-term research recommended by the task group would probably be carried out 10 years or more in the future but might require some near-term planning. This recommended research is ranked in terms of its potential for expanding knowledge of carbon compounds in the solar system and for its close relationship to research and missions currently in progress or recently completed.

Titan. Titan is believed to be a major reservoir of organic materials in the solar system, and the dynamic processes of Titan's atmospheric chemistry provide an ongoing example of the abiotic formation of complex organics from methane. This satellite merits close scrutiny by continued ground-based observation and computer and laboratory modeling of its atmospheric chemistry.

Recommendation: Planning should start now for a follow-up of the Cassini mission to Titan that would include a lander sent to sample its surface, since the complexity of the organics there is expected to be much greater than that of the organics in its atmosphere. The lander should have the capability of sampling organic materials that are solids at 96 K as well as those that are liquids. The Titan Explorer mission considered by the solar system exploration decadal survey is a good starting point for this planning.

Primitive Bodies. The successful landing of the NEAR spacecraft on the asteroid Eros has demonstrated the feasibility of sending a probe to an asteroid. The solar system exploration decadal survey report recommended in situ and sample-return missions to asteroids and comets to provide direct information on the structures of the organic compounds present in comets and asteroids and to provide information about whether or not the asteroids are the sources of meteorites and dust reaching Earth.

Recommendation: In situ analyses as well as sample-return missions should be performed for both asteroids and comets. The task group points to the solar system exploration decadal survey report's recommended New Frontiers-class Comet Surface Sample Return mission[4] as an example of an activity that would greatly enhance understanding of the organic constituents of the solar system's primitive bodies.

Current and upcoming missions are targeted to active comets; Pluto/Charon and perhaps one or two Kuiper Belt objects; and asteroids of spectral classes S, G, and V. Most of these missions have not been optimized for the study of organic materials even though the population of primitive small bodies may preserve organic materials from a wide range of nebular heliocentric distances. A rich research opportunity exists to explore these different chemical and thermal regimes, thus enabling an understanding of the distribution and history of organic materials in the solar system.

Recommendation: Every opportunity should be taken to direct space missions to small bodies to do infrared spectral studies of these targets, especially a D- or P-type asteroid, to determine if these dark bodies contain an appreciable amount of carbon compounds and, if so, whether they are the sources of the carbonaceous meteorites and dust reaching Earth.

In this regard, a possible opportunity is conducting such studies as an adjunct to the Trojan Asteroid/Centaur Reconnaissance flyby mission described in the solar system exploration decadal survey.[5] Although this mission was not ranked in the survey's final list of priorities, the possibility of using a single spacecraft to make a sequential flyby of three different classes of primitive bodies—i.e., a D- or P-type main-belt asteroid, a jovian Trojan asteroid, and a Centaur—has sufficient merit to warrant additional study for possible implementation as a New Frontiers mission at some time in the future.

Europa, Callisto, and Ganymede. In the early 2000s, NASA's solar system exploration plans included a Europa Orbiter mission that would undertake flyby observations of Callisto and Ganymede prior to entering orbit about Europa. Although excessive cost growth led to the cancellation of this mission, scientific interest in the study of Jupiter's large, icy satellites continues to be strong. The Europa Geophysical Explorer, a somewhat more elaborate version of the Europa Orbiter, was the highest-priority large mission recommended by the 2003 solar system exploration decadal survey.[6] NASA responded to the survey's recommendation by initiating the development of the Jupiter Icy Moons Orbiter (JIMO) mission, the first of a line of advanced-technology spacecraft with significantly expanded science capabilities compared to previous concepts for missions to Europa. JIMO would have conducted global mapping of all three icy satellites, at resolutions of 10 m or better, and might have included a small Europa lander. Organic materials can be studied by making provisions for high-signal-to-noise-ratio spectroscopy at resolutions adequate to discriminate potential carbon-bearing species in both high- and low-albedo regions. JIMO was indefinitely deferred in 2005, and NASA and the planetary science community are currently assessing plans for a more conventional and very much less expensive alternative.[7]

Recommendation: The task group reiterates the solar system exploration decadal survey's findings and conclusions with respect to the exploration of Europa and recommends that NASA and the space science community develop a strategy for the development of a capable Europa orbiter mission and that such a mission be launched as soon as it is financially and programmatically feasible. Any future Europa lander mission should be equipped with a mass spectrometer capable of identifying simple organic materials in a background of water and hydrated silicates.

NOTES

1. National Research Council (NRC), *New Frontiers in the Solar System: An Integrated Exploration Strategy*, The National Academies Press, Washington, D.C., 2003, pp. 206-207.
2. NRC, *New Frontiers in the Solar System: An Integrated Exploration Strategy*, 2003.
3. NRC, *New Frontiers in the Solar System: An Integrated Exploration Strategy*, 2003, pp. 198-200.

    4. NRC, *New Frontiers in the Solar System: An Integrated Exploration Strategy*, 2003, p. 195.
    5. NRC, *New Frontiers in the Solar System: An Integrated Exploration Strategy*, 2003, p. 25.
    6. NRC, *New Frontiers in the Solar System: An Integrated Exploration Strategy*, 2003, p. 4.
    7. NRC, *Priorities in Space Science Enabled by Nuclear Power and Propulsion*, The National Academies Press, Washington, D.C., 2006, pp. 17-20.

I—The Chemistry of Carbon

The chemistry of carbon is richly variable and serves as the basis for life on Earth. Away from Earth, in the solar system and beyond, carbon and its compounds are abundant. Indeed, the range of chemical structures found on and off Earth are virtually countless. The variations in structure provide evidence of conditions that prevailed at times before Earth was formed and at cosmic distances. They record events that occurred during the formation of the solar system and during the division of materials among planets. In that sense, variations in the chemical structures of carbon compounds are some of our ultimate historical documents. The exploration and decoding of the carbon-chemical record are providing some of the best information about relationships between our planet, the solar system, and the cosmos. A major challenge, however, is to be able to differentiate biotic from abiotic molecules.

1

Biotic and Abiotic Carbon Compounds

CARBON COMPOUNDS—DEFINITIONS AND CHARACTERISTICS

Historically, chemists have referred to all compounds of carbon except the oxides, carbonates (e.g., limestone and marble), metallic carbides, and elemental forms (e.g., diamond and graphite) as "organic." But, despite the term, organic molecules, even elaborate ones, are not necessarily produced biotically. Discussed below are several important characteristics of organic molecules that, as we shall see, are pertinent to the problem of trying to differentiate between biotic and abiotic origin:

1. Structural stability over wide ranges of temperature and pressure;
2. Aromaticity;
3. Aliphatic homologs;
4. Stereoisomerism; and
5. Structural complexity.

These characteristics are discussed in detail in subsequent sections.

Structural Stability

Carbon atoms can bond strongly to each other. Chains of atoms can extend indefinitely and, because each atom can form four bonds, with multiple branches. Chains can bend around to form rings, and rings can be fused to form sheets of atoms for which the bonding diagrams look like chicken wire. In such cases, the four bonds at each carbon atom are often arranged to provide double or even triple linkages (see the next section, "Aromaticity"). Such materials—and many smaller, simpler molecules—are capable of outliving our planet. If they are tucked away in some bit of space rock or ice, they are literally waiting to be discovered and interrogated. The chemical structure of a carbon-containing molecule (the bonding pattern and the other chemical elements that are present) provides information about the last time that molecule was warm enough to rearrange spontaneously or to react with another molecule. Depending on the molecule in question, "warm" might mean some temperature more than a hundred degrees below water's freezing point or some temperature above the melting point of metallic zinc. The conditions monitored would be as follows:

- What other chemical elements and compounds were present?
- What was the pressure?
- What mineral surfaces were available?
- What sort of electromagnetic radiation was present?

Knowing the limits on conditions could provide clues that might reveal a biotic or an abiotic origin for compounds sampled.

Aromaticity

One of the most important characteristics of organic compounds is their ability to form molecules with a distinctive stability resulting from aromaticity, with the electrons of neighboring carbon atoms shared "in resonance." An aromatic compound is one in which at least two equivalent structures exist. Benzene is the simplest example (Figure 1.1).

Aromatic compounds are particularly unreactive, or stable, and despite the name are not necessarily volatile. Benzene rings are easily elaborated into more complex, polycyclic structures by the one-ring build-up mechanism.[1] Polycyclic aromatic hydrocarbons (PAHs) are common structural components of dyes, soot from combustion, and interstellar particles. Some simple examples are shown in Figure 1.2. PAHs in nature, particularly those in interstellar particles, can include hundreds of benzene rings fused so that most share all of their sides.

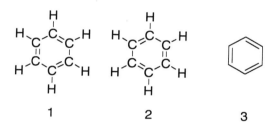

1   2   3

FIGURE 1.1 Three alternative views of the chemical structure of benzene, C_6H_6. In all structure diagrams, the lines represent chemical bonds. The letters C and H represent atoms of carbon and hydrogen, respectively. Structures 1 and 2 are identical except that the double bonds between the carbon atoms are rearranged. Because the locations of the atoms are unchanged even though the bonds have moved, the structures are said to be "in resonance," and the *actual* structure of benzene is understood to be a combination of these two representations. Structure 3 is an organic-chemical-shorthand version equivalent to structures 1 and 2. In such shorthand structures, bonds to H are omitted. Each carbon is known to have four bonds and, wherever one is missing, it is understood that an unseen bond to H is present. This convention removes clutter from the drawings. Wherever a bond ends or two bonds meet and no elemental symbol is shown, it is understood that a carbon atom is present.

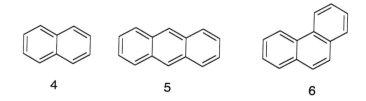

4   5   6

FIGURE 1.2 Shorthand representations (see Figure 1.1) of the three simplest polycyclic aromatic hydrocarbons, respectively naphthalene (4), anthracene (5), and phenanthrene (6).

Aliphatic Homologs

A comment should be made to contrast aromatic with aliphatic organic compounds. Aliphatic hydrocarbons (also called paraffins) are straight-chain alkanes (i.e., compounds having the general formula C_nH_{2n+2}, where n is an integer greater than one) starting with methane (CH_4) and increasing by $-CH_2-$ units as a homologous series to high molecular weights. They can form abiotically (CH_4 to typically decane, under hydrous conditions even to $> C_{35}$)[2] and biotically by direct synthesis or by geological degradation of lipid and cell membrane detritus.[3] Lipids are aliphatic homologous compounds (e.g., fatty acids, fatty alcohols, etc.) important as membrane components and for energy storage. They are currently biologically synthesized but can also form abiotically in aqueous media at elevated temperatures and pressures.[4-6] Biomarkers are organic compounds with specific structures (e.g., cholestane) that can be related back to their natural product precursors. The natural products (e.g., cholesterol) are biosynthesized from lipid precursors.[7] Homologous aliphatic organic compounds and biomarkers can be distinguished by organic geochemists as being derived from abiotic or biotic sources.[8] The homologous compounds with the biomarkers should be considered as the intermediary organics between the CH_4 chemistry on planetary bodies and the formation of aromatics in the solar and interstellar medium as additional suitable tracers for evidence of life.

Stereoisomerism

Molecules are three-dimensional. A carbon atom with four single bonds lies at the center of a tetrahedron. The atoms to which the carbon is bonded are at the vertices of the tetrahedron. Those atoms are in turn likely to be bonded to other atoms. If the chemical structures at the four corners all differ, however slightly, the mirror images of the tetrahedron will not be superimposable (Figure 1.3).

Mirror-image stereoisomers that are not superimposable are called enantiomers; stereoisomers that are not enantiomers are called diastereomers. Enantiomers possess chirality, or handedness, and when dissolved rotate the plane of polarized light when it is passed through the solution.

Life on Earth makes use of only a limited number of diastereomers of all those that are possible.[9] Moreover, biotic processes display an enantiomeric excess; e.g., left-handed amino acids and right-handed sugars almost exclusively predominate in living systems.

Carbon atoms bearing four different substituents are said to be chiral centers. If a molecule has n chiral centers it will, in most cases, have 2^n stereoisomers. There will, for example, be 256 stereoisomers of a compound with eight chiral centers. Each will have exactly the same chemical formula and pattern of connectivity among its atoms (A is connected to B is connected to C and D, and so on). Only the arrangements of those atoms in space will differ, and there will be 256 variations. Life functions by using only a small subset of all possible stereoisomers.

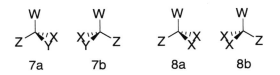

FIGURE 1.3 An illustration of stereoisomerism. In these depictions of tetrahedral carbon atoms, bonds represented by straight lines lie in the plane of the paper. Those represented by wedges project in front of the paper (the filled wedges) or to the rear (the broken wedges). W, X, Y, and Z represent different chemical groups, anything from a single atom (an H, for example) to a complex chemical substituent with many atoms in addition to the one that is bonded directly to the carbon atom. In structure 7, all four groups are different. The mirror images, 7a and 7b, cannot be rotated so that the structures are superimposable. In structure 8, by contrast, two of the groups are identical. If structure 8b is rotated 180° about its vertical axis, it can be superimposed on 8a (i.e., it is seen to be identical to 8a). Mirror images of tetrahedra will be nonsuperimposable only when all four vertices are different.

Structural Complexity and Chemical Reactivity

Descriptions of organic chemical structures focus on two distinct topics: a molecule's "carbon skeleton" and its "functional groups" (if any). The first of these terms is practically self-defining. The carbon skeleton, a framework of chemical bonds, is usually the most important determinant of a molecule's size and shape. The functional groups are chemical ornaments attached to that skeleton. They are usually the most important determinants of a molecule's reactivity. Examples follow.

Carbon skeletons are best described by structural formulas like those shown in Figures 1.1 through 1.3. If the structure contains no double bonds or rings, seven carbon atoms can be assembled to produce the nine distinct carbon skeletons shown in Figure 1.4.

As drawn, the structural formulas in Figure 1.4 refer specifically to the molecule C_7H_{16}. As indicated by the elemental formula, which indicates that only carbon and hydrogen are present, it is a hydrocarbon, unadorned by functional groups. Numbers of possible structures for such compounds increase rapidly with the number of carbon atoms. The molecular formula $C_{10}H_{22}$ leads to 75 distinct structures. Doubling the number of carbon atoms (i.e., to $C_{20}H_{42}$) makes the total 366,319. Another doubling, to $C_{40}H_{82}$, yields 62,481,801,147,341 possible structures. However, as discussed above for stereoisomers, life functions by using only a small subset of all possible structures, patterns, and isomers.[10] For example, a nonregular distribution of alkanes (e.g., an imbalance of even versus odd numbers of carbon atoms) and mainly linear not branched isomers occur in living systems.

The number of possible structures increases even more rapidly when multiple bonding and ring formation are considered. As an example, Figure 1.5 shows the structural formulas for the C_7 hydrocarbons containing a five-membered ring and up to one double bond.

Many further structures could be drawn for C_7 hydrocarbons. These would include 3-, 4-, 6-, or 7-membered rings and double and triple bonds limited only by the requirement that each carbon atom always have four bonds.

Functional groups contain elements in addition to carbon and hydrogen. The resulting polarity and presence of nonbonding electrons make them likely sites for chemical reactions. Examples are shown in Figure 1.6.

It is not practical here to review in detail the chemistry of functional groups, but the basic factors are easy to appreciate. Nitrogen and oxygen contain respectively one and two more valence electrons than does carbon. Nitrogen has to make only three chemical bonds to complete its valence octet. Oxygen can only make two. Both nitrogen and oxygen are more electronegative than carbon. C-N and C-O bonds are therefore polar. The carbon within them is susceptible to attack by nucleophiles (chemical reactants seeking to stabilize high electron density by forming a chemical bond to an electron-poor site, such as the C in a C=O bond). Simultaneously, the nonbonding electron pairs on nitrogen and oxygen (two pairs in that case) are available to participate in electronic rearrangements.

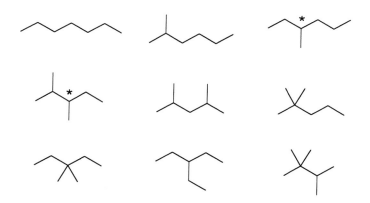

FIGURE 1.4 Shorthand representations (see caption for Figure 1.1) of the nine different carbon skeletons possible for a molecule with seven carbon atoms and no double bonds or rings. The asterisks mark chiral centers (see Figure 1.3). Each of the structures that contain a chiral center will exist as two stereoisomers.

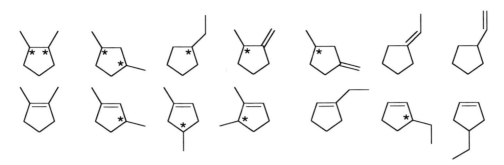

FIGURE 1.5 Structural formulas for hydrocarbons containing seven carbon atoms, one five-membered ring, and up to one double bond. The asterisks mark chiral centers (see Figure 1.3). Both of the structures containing two chiral centers happen also to contain a plane of symmetry. As a result, two of the stereoisomers will be equivalent and the molecules in question will have only three stereoisomers instead of $2^2 = 4$.

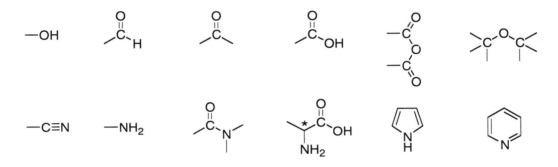

FIGURE 1.6 Functional groups. The open bonds represent points of attachment to carbon skeletons. From left to right in the top row, these examples include an alcohol, an aldehyde, a ketone, a carboxylic acid, a carboxylic acid anhydride, and an ether. Nitrogen-containing functional groups are shown in the second row. From left to right these include a nitrile (or organic cyanide), an amine, an amide, an α-amino acid, pyrrole, and pyridine.

Commonly, two carbon skeletons can be, in essence, glued together when the functional group on one attacks a functional group on the other and, as a result, a chemical bond is formed. An example is shown in Figure 1.7. In this case an alcohol with a carbon skeleton similar to those shown in Figure 1.5 reacts with a carboxylic acid having a carbon skeleton similar to one of those shown in Figure 1.4. The products are a molecule of water and an organic molecule in which the two carbon skeletons are connected by an ester linkage. Such reactions, in which two smaller molecules combine to form a larger product with the release of some small molecule, are termed "condensation reactions."

Polymers exemplify both structural complexity (or, at least, great molecular size) and chemical reactivity. Polymers are synthesized from one or more monomers. Often, the monomers have functional groups that allow them to link their carbon skeletons together to form endless chains (addition polymers like polyethylene are the exception). Dacron and nylon are familiar polymers. The former is a polyester containing linkages like that illustrated in Figure 1.7. The latter is a polyamide (the amide structure is shown in Figure 1.6). The pertinent monomers and reactions are summarized in Figure 1.8.

Polymers are macromolecules. Natural products that are polymeric macromolecules include proteins (monomers = amino acids), nucleic acids (monomers = nucleotides), and cellulose (monomer = glucose). Coal is a natural macromolecule but does not qualify as a polymer because it has no regular, repeating structure based on a restricted set of monomers. Instead, it is a product of the condensation of cellulose, lignin, and other plant and bacterial products in the anaerobic, fermentative systems described as "coal swamps." Another nonpolymeric

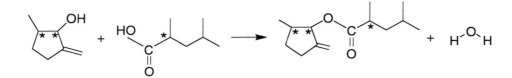

FIGURE 1.7 A typical condensation reaction. The alcohol and carboxylic acid shown as reactants are examples of carbon skeletons similar to those introduced in Figures 1.4 and 1.5 bearing two of the functional groups shown in Figure 1.6. For each carbon skeleton, the addition of the functional group has created a new chiral center by introducing asymmetry at a formerly symmetrical carbon position. In the absence of stereochemical control, the reacting alcohol will exist as a mixture of $2^2 = 4$ stereoisomers, the carboxylic acid will be a mixture of 2 stereoisomers, and the organic product, an ester, will be a mixture of $2^3 = 8$ stereoisomers.

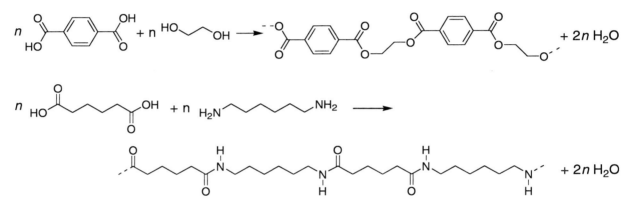

FIGURE 1.8. Chemical reactions depicting the formation of two familiar polymers, namely Dacron (top) and nylon (bottom). The dashed bonds indicate that the repeating, polymeric structures extend indefinitely. Values of n can exceed 10^4. The monomers used to form Dacron are a benzene dicarboxylic acid and a C_2 dialcohol. The product is a polyester. Nylon is formed by polymerizing two C_6 monomers, a dicarboxylic acid and a diamine. The product is a polyamide. Both polymerizations are condensation reactions. A molecule of water is released for each ester or amide link that is formed.

natural macromolecule, kerogen, forms from the condensation of algal and microbial debris in seafloor muds. The depositional environment (i.e., constant deliveries of silt, clay, lime, and other inorganic debris; ongoing metabolism of the organic matter) is such that kerogen seldom constitutes more than 10 percent of the mass of even a carbon-rich sedimentary rock like a petroleum "source rock." Kerogen is nevertheless a familiar object of study for organic geochemists. It attracts attention because of the tendency of organic compounds to react with each other, it often constitutes 99 percent of the organic carbon in a sedimentary rock.

Kerogen, a macromolecule so large that it is not soluble in any solvent, is isolated by dissolving everything else. A sedimentary rock is ground to produce a powder. Small organic molecules are extracted using organic solvent. Inorganic material, the rocky matrix, is dissolved using hydrochloric and hydrofluoric acids. The kerogen remains. Ideas about its molecular structure (which will vary depending on the specific algal and microbial precursors and conditions, such as the abundance of H_2S, in the depositional environment) can be obtained by chemical and thermal degradation of the macromolecule.

Insoluble, macromolecular, carbonaceous debris will form wherever reactive organic molecules are stored in close proximity. A terrestrial example of such material is kerogen. An extraterrestrial counterpart is found in carbon-rich meteorites. It may be called "kerogen" in written reports, but the usage is purely operational and does not require a biotic origin: The material is insoluble organic carbon.

DETERMINATION OF MOLECULAR ORIGINS

The exploration of organic cosmochemistry is not a search for life but an examination of *all* of the processes that have shaped the existence of life on Earth and, perhaps, elsewhere in the universe. The keys are to look in the right places, to collect and examine the right molecular mixtures, and to patiently and systematically extend the organic chemical analyses in ways that will maximize the precision with which interpretations can be made.

Each organic molecule provides information about its origins through details of its structure. Within the carbon skeleton, the number of rings and multiple bonds (which is to say the ratio of hydrogen to carbon, since each ring, double bond, or triple bond decreases the number of bonds to hydrogen) provides information about the availability of hydrogen at the site of molecular synthesis. Structures rich in multiple bonds suggest that reactive hydrogen was rare during assembly of the carbon skeleton. Some carbon skeletons, for example those containing small (3- or 4-membered) rings, incorporate quite a bit of strain (i.e., the angles between bonds are smaller than usual). If such rings have survived, they suggest that the molecule was unable to rearrange as it was formed and that the molecule has been cool since the time of its synthesis. Similarly, biotic processes synthesize many unique chemical structures.[11] Such lines of evidence ramify almost endlessly, being limited only by the level of detail to which the structure is known.

Functional groups provide additional evidence about a molecule's history. Most obviously, the presence of nitrogen and/or oxygen indicates that reactive forms of those elements must have been available at the site of molecular synthesis. The nature of the functional group carries further information. For example, alcohols, aldehydes, and carboxylic acids indicate progressively higher levels of oxidation. The survival of functional groups that would react with water (e.g., acid anhydrides or nitriles) indicates that the environments of synthesis and storage were dry. Again, the extension and elaboration of interpretations are limited only by the quality of the information about the numbers and types of functional groups.

Within mixtures of molecules, discernible patterns can be interpreted in terms of controlling mechanisms. A preference for molecules with even numbers of carbon atoms, for example, would indicate that C_2 reactive units had been important in the environment of synthesis. A prevalence of two-dimensional, sheet-like structures would suggest formation on surfaces. In general, the properties of catalysts can be discerned from the extent to which some products have been favored over others. Where catalysis has been controlled precisely, the structural preferences can be absolute. At this, life, i.e., biochemistry, is the champion.

The common, biologically synthesized amino acids, the monomers used in the synthesis of proteins, provide an illustrative example. By definition, an amino acid contains both an amine group and a carboxylic acid group (see Figure 1.6). The carbon skeletons of the amino acids used in proteins contain up to 10 carbon atoms and up to six double bonds or rings. With two different functional groups, 10 carbon atoms, and varying numbers of rings and double bonds, the number of possible molecular structures is astronomical. From among those structures, each cell makes only 20 for inclusion in proteins. Each contains the particular assembly of functional groups identified in Figure 1.6 as an α-amino acid. As indicated in Figure 1.6, if the open bond is attached to anything but another H (the carbon shown as having only three bonds must have an unseen bond to H), another NH_2, or another CO_2H, the molecule will contain at least one chiral carbon. At that point, the biosynthetic process is even stereoselective, producing only one of the stereoisomers of each amino acid.

Encountering that mixture of materials, with only 20 structures from among the possible millions and only one chiral form of each, anyone would recognize the catalyst as being a biological process. Abiotic processes are less precisely selective, but the logical process of working backward from the analysis of an organic mixture to inferences about the mechanism of synthesis is no different.

Molecules also carry information about their origins in one rather unconventional way. The elements that are most important in organic chemistry—carbon, hydrogen, nitrogen, oxygen, and sulfur—all happen to have multiple stable isotopes (specifically, ^1H and ^2H; ^{12}C and ^{13}C; ^{14}N and ^{15}N; ^{16}O, ^{17}O, and ^{18}O; and ^{32}S, ^{33}S, ^{34}S, and ^{36}S). Even on Earth, the relative abundances of those isotopes vary measurably (one part in 10^3 or, as it is usually written, 1‰) as a result of isotope effects that are associated with natural processes ranging from the evaporation of water to the fixation of carbon dioxide by photosynthetic organisms. Isotope effects associated with industrial processes such as the cracking of petroleum or electrolytic production of H_2 from H_2O also fractionate the stable isotopes. In the cosmos, isotopic abundances vary not only as a result of isotope effects but also because of variations in nucleosynthetic processes.

Isotopic evidence can be used in two general ways. First, mixtures of compounds from different sources can be recognized. Compounds in one population of molecules might, for example, all be depleted in ^{13}C. Another population might be distinguished by enrichment in ^2H. In favorable cases, two or more populations can be recognized and their isotopic characteristics defined well enough that, when a particular compound happens to have multiple origins, mixing equations can be used to determine the proportion attributable to each source. Second, precursor-product and other genetic relationships can be recognized. If compound A is consistently depleted in ^{13}C relative to compound B by a few parts per thousand, it is likely that A derives from B and that the reaction relating A and B has a kinetic isotope effect. If one family of structurally related compounds has isotopic abundances that are similar or that covary systematically, it is likely that all members of the family share a common source, and if another compound is related structurally but does not conform to the isotopic pattern, it must have a separate origin in spite of the structural relationship.

CRITERIA FOR DISTINGUISHING BETWEEN BIOTIC AND ABIOTIC COMPOUNDS

The chemistry of carbon leads to extraordinary variations in the structures of organic molecules. Factors influencing those variations include the following:

- The abundances of reactive forms of some of the most abundant elements in the cosmos, namely hydrogen, nitrogen, and oxygen;
- The temperatures and pressures at which the organic compounds have formed;
- The temperatures and times of storage of the organic chemical products; and
- The nature of any catalysts—ranging from random solid surfaces to other organic molecules—that were present in the environment of synthesis.

Knowledge of organic chemistry is so advanced, and the information content of mixtures of organic compounds (i.e., that encoded by variations in molecular compositions and structures) is so great, that inverse problems can be attacked with considerable success. Given the structures and relative abundances of organic compounds from samples of cosmochemical interest, it may be possible to work backward and to reconstruct the physical and chemical conditions prevailing at the times of their synthesis and throughout their subsequent history.

With regard to the indicators that might differentiate between a biotic and an abiotic origin for particular organic compounds, the task group found that the most compelling indicators of an abiotic origin include the following:

- The presence of a smooth distribution of organic compounds in a sample, e.g., a balance of even versus odd numbers of carbon atoms in alkanes;
- The presence of all possible structures, patterns, isomers, and stereoisomers in a subset of compounds such as amino acids;
- A balance of observed entantiomers; and
- The lack of depletions or enrichments of certain isotopes with respect to the isotopic ratio normally expected.

Likewise, the converse of the above items are indicators of possible biotic synthesis. Thus, the following are indicators of a biotic origin:

- The presence of an irregular distribution of organic compounds in a sample, e.g., an imbalance of even versus odd numbers of carbon atoms in alkanes;
- The presence of only a small subset of all possible structures, patterns, isomers, and stereoisomers;

- An imbalance of observed entantiomers; and
- The depletion or enrichment of certain isotopes with respect to the isotopic ratio normally expected.

However, some abiotic processes can mimic biotic ones and vice versa, and inferences will necessarily be based on several indicators and will, of course, be probabilistic.

NOTES

1. B.R.T. Simoneit and J.C. Fetzer, "High Molecular Weight Polycyclic Aromatic Hydrocarbons in Hydrothermal Petroleums from the Gulf of California and Northeast Pacific Ocean," *Organic Geochemistry* 24: 1065-1077, 1996.

2. A.I. Rushdi and B.R.T. Simoneit, "Abiotic Condensation Synthesis of Glyceride Lipids and Wax Esters Under Simulated Hydrothermal Conditions," *Origins of Life and Evolution of Biospheres* 36: 93-108, 2006.

3. See, for example, B.R.T. Simoneit, "Biomarkers (Molecular Fossils) as Geochemical Indicators of Life," *Advances in Space Research* 33: 1255-1261, 2004.

4. A.I. Rushdi and B.R.T. Simoneit, "Lipid Formation by Aqueous Fischer-Tropsch-type Synthesis over a Temperature Range of 100-400° C," *Origins of Life and Evolution of the Biosphere* 31: 103-118, 2001.

5. A.I. Rushdi and B.R.T. Simoneit, "Condensation Reactions and Formation of Amides, Esters, and Nitriles Under Hydrothermal Conditions," *Astrobiology* 4: 211-224, 2004.

6. A.I. Rushdi and B.R.T. Simoneit, "Abiotic Condensation Synthesis of Glyceride Lipids and Wax Esters Under Simulated Hydrothermal Conditions," *Origins of Life and Evolution of Biospheres* 36: 93-108, 2006.

7. B.R.T. Simoneit, "Biomarkers (Molecular Fossils) as Geochemical Indicators of Life," *Advances in Space Research* 33: 1255-1261, 2004.

8. See, for example, B.R.T. Simoneit, "Biomarkers (Molecular Fossils) as Geochemical Indicators of Life," *Advances in Space Research* 33: 1255-1261, 2004.

9. Roger Summons, Department of Earth, Atmospheric, and Planetary Sciences, Massachusetts Institute of Technology, "Molecular Biosignatures: Real and Potential Biomarkers, Analytical Innovations, Meteorites, and Mars," presentation to the Committee on the Origins and Evolution of Life, January 25, 2006, Beckman Center, Irvine, California.

10. Summons, "Molecular Biosignatures," presentation January 25, 2006.

11. See, for example, B.R.T. Simoneit, "Biomarkers (Molecular Fossils) as Geochemical Indicators of Life," *Advances in Space Research* 33: 1255-1261, 2004.

II—The Formation, Modification, and Preservation of Organic Compounds in the Solar System

The search for understanding of how organic environments originated on the early Earth and throughout the universe and their association with life processes is the ultimate interdisciplinary field. Subdisciplines of the Earth sciences, biology, chemistry, astronomy, and the space sciences are all needed to contribute to the contemporary understanding of this complex problem.

The search for organics in the solar system involves a series of interrelated questions. These include where organic reservoirs are located in the solar system,[1] what specific organic compounds are present in these reservoirs, and what the location and identification of these organic compounds can tell us about both the evolution of the solar system and the possible presence of life at locations other than on Earth.

The discussion in Part II serves as an overview of all known inventories of organic compounds in the solar system, the possible means by which they were formed, and also—based on current observations—locations, not yet examined in detail, where organic compounds may be present. Recommendations for further exploration take into consideration the likelihood that significant organics will be found, the ease with which they can be found, and the anticipated amount and significance of data that can be accumulated from a particular research activity or spacecraft mission.

All the carbon in the universe is made by fusion reactions in stars. Carbon-12 (^{12}C) is created by the fusion of three helium-4 (^{4}He) nuclei. Carbon-13 (^{13}C) is made late in the lives of red giant stars, where formation of helium, catalyzed by ^{12}C, results in the formation of ^{13}C, ^{14}N, and ^{15}O via the CNO cycle outlined below:

$$^{12}C + {}^1H \rightarrow {}^{13}N + \gamma$$
$$^{13}N \rightarrow {}^{13}C + e^+ + \nu$$
$$^{13}C + {}^1H \rightarrow {}^{14}N + \gamma$$
$$^{14}N + {}^1H \rightarrow {}^{15}O + \gamma$$
$$^{15}O \rightarrow {}^{15}N + e^+ + \nu$$
$$^{15}N + {}^1H \rightarrow {}^{12}C + {}^4He$$

---

[1] For the purposes of this report, the interstellar medium is included in this search, both because Earth-based observations can be used to identify organics present there, and because it is thought that the interstellar medium was a source for organics present in the solar system's protoplanetary disk.

where e^+, γ, and ν are a positron, a gamma ray, and a neutrino, respectively. The $^{12}C/^{13}C$ ratio resulting from the CNO cycle is in the range of 15 to 20.

The CNO cycle proceeds in the shells of the red giant stars that contain high levels of carbon, where the conversion of ^{12}C to ^{13}C, and to ^{14}N and ^{15}O, proceeds after the bulk of the hydrogen has been converted to He. The pulsation of the red giant during this energetic process disperses the elements formed into the interstellar medium, where they serve as the starting materials for the formation of a new star.

The first organic compounds were formed from the carbon injected into the interstellar medium under the influence of cosmic rays and ultraviolet light. Simple hydrocarbons and other compounds that contain nitrogen, oxygen, and sulfur were formed in this cloud of dust and molecules. This process proceeded for about 10^7 years, producing additional organics before the dust cloud collapsed to form stars and their associated planetary systems.

In the solar system the evolution of carbon compounds proceeded during planetary system formation. The existing compounds were subjected to the shock waves resulting from the collapse of the dust cloud to stars and protoplanetary disks. The intense ultraviolet and x-rays emitted by the new star effected changes in some of the organics. Carbon compounds ultimately derived from the interstellar medium were accreted onto planetesimals in the early solar system, where considerable thermal and aqueous modification may have occurred. These planetesimals then aggregated to form planets, a process that further modified some of their organic constituents. The organics present in the atmospheres of the newly formed planets were subjected to solar ultraviolet radiation as well. Organics on and below the surface of planets were further changed by energy sources including heat from volcanoes, heating by transport into planetary interiors where they were subjected to heat and pressure, contact with hydrothermal systems that initiated reactions with water at high temperatures and pressure, and reduction by minerals. Volcanoes also injected volatile organics into the atmosphere where solar ultraviolet radiation and x-rays changed them.

2

Interstellar Chemistry

THE INTERSTELLAR MEDIUM

Inventory of Organic Compounds in the Interstellar Medium

Some 20 to 30 percent of the mass of our galaxy is in the form of the interstellar medium (ISM), i.e., the material between the stars. The ISM consists primarily of gas, with atomic or molecular hydrogen and helium contributing approximately two-thirds and one-third of the total mass, respectively. The next most abundant atoms, oxygen, carbon, and nitrogen, collectively account for about 1 percent of the ISM's mass. The remaining elements are present only in trace amounts. Approximately 1 percent of the mass of the ISM is in the form of micron-size dust particles. Astronomical observations, combined with studies of interstellar grains preserved in meteorites, suggest that the dust might consist variously of amorphous carbon, complex fullerenes, polycyclic aromatic hydrocarbons, diamond, silicon carbide, silicates, carbonates, and a host of other candidates, all with or without mantles of ices and/or organic compounds.[1]

Important components of the ISM are molecular clouds, which are dense, massive objects found throughout the Milky Way and in many external galaxies. In these molecular clouds—also known as dense clouds—the gas density is 10^3 to 10^6 particles/cm^3, which is very high by interstellar standards, and their masses can be as large as a million times the mass of the Sun. They are also usually very cold objects, with temperatures typically in the range from 10 to 100 K. Because of their high masses, these objects are the sites of star and planet formation, and also where complex gas-phase chemistry occurs.

Astronomical observations of the ISM have revealed the presence of numerous organic compounds. More than 125 different chemical species have been identified in interstellar and circumstellar regions, some containing 10 or more carbon atoms (Table 2.1). Assuming that the carbon in the ISM is present in cosmic abundance, then only 0.04 percent (by number) of the material there is carbon, even though approximately 80 percent of the observed species in the ISM are organic, including almost all of the larger molecules, many of which are relatively complex.[2] Organic compounds are, however, only a trace constituent of the ISM and account for less than 1 percent of its total mass. Inorganic compounds abound, with CO, for example, accounting for some 20 percent of the carbon in dense interstellar clouds. CO is, itself, outnumbered by the most common molecular species, H_2, by a factor of approximately 10,000. The majority of the molecular species identified in the ISM have been discovered using high-resolution (1 part in 10^6 to 10^8) spectroscopic techniques of radio and millimeter astronomy. This

TABLE 2.1 Known Interstellar and Circumstellar Molecules

2	3	4	5	6	7	8	9	10	11	12	13
									Number of Atoms[a]		
H_2	C_3	$c\text{-}C_3H$	C_5	C_5H	C_6H	CH_3C_3N	CH_3C_4H	CH_3C_5N?	HC_9N	$CH_3OC_2H_5$	$HC_{11}N$
AlF	C_2H	$l\text{-}C_3H$	C_4H	$l\text{-}H_2C_4$	CH_2CHCN	$HCOOCH_3$	CH_3CH_2CN	$(CH_3)_2CO$			
$AlCl$	C_2O	C_3N	C_4Si	C_2H_4	CH_3C_2H	CH_3COOH?	$(CH_3)_2O$	NH_2CH_2COOH?			
C_2	C_2S	C_3O	$l\text{-}C_3H_2$	CH_3CN	HC_5N	C_7H	CH_3CH_2OH	CH_3CH_2CHO			
CH	CH_2	C_3S	$c\text{-}C_3H_2$	CH_3NC	$HCOCH_3$	H_2C_6	HC_7N				
CH^+	HCN	C_2H_2	CH_2CN	CH_3OH	NH_2CH_3	CH_2OHCHO	C_8H				
CN	HCO	CH_2D^+?	CH_4	CH_3SH	$c\text{-}C_2H_4O$	CH_2CHCHO					
CO	HCO^+	$HCCN$	HC_3N	HC_3NH^+	CH_2CHOH						
CO^+	HCS^+	$HCNH^+$	HC_2NC	HC_2CHO							
CP	HOC^+	$HNCO$	$HCOOH$	NH_2CHO							
CSi	H_2O	$HNCS$	H_2CHN	C_5N							
HCl	H_2S	$HOCO^+$	H_2C_2O	HC_4N							
KCl	HNC	H_2CO	H_2NCN								
NH	HNO	H_2CN	HNC_3								
NO	$MgCN$	H_2CS	SiH_4								
NS	$MgNC$	H_3O^+	H_2COH^+								
$NaCl$	N_2H^+	NH_3									
OH	N_2O	SiC_3									
PN	$NaCN$	C_4									
SO	OCS										
SO^+	SO_2										
SiN	$c\text{-}SiC_2$										
SiO	CO_2										
SiS	NH_2										
CS	H_3^+										
HF	$SiCN$										
SH	$AlNC$										
FeO(?)	$SiNC$										

NOTE: The observations are of molecular emission by high-resolution spectroscopy with lines having a quality factor of 1 part in 10^6 to 10^8. The identification of these molecules has been made on the basis of their pure rotational, rovibrational, or electronic spectra, which occur in the radio/millimeter, infrared, and optical/ultraviolet regions of the electromagnetic spectrum, respectively. The species are a composite of data obtained from a variety of astronomical sources, including comets, several dense interstellar clouds, and circumstellar envelopes.

[a]A lower-case "c" indicates a cyclic structure; a lower-case "l," a linear structure; and a "?," a tentative identification.

SOURCE: Courtesy of H. Alwyn Wootten, National Radio Astronomy Observatory, available at http://www.cv.nrao.edu/~awootten/allmols.html, last accessed January 22, 2007.

triumph of radio astronomy has changed our perception of the universe as being predominantly a rarified atomic environment to one containing a large (organic) molecular component.

The Synthesis of Interstellar Molecules

The observed interstellar species listed in Table 2.1 are not those expected based on conditions of thermodynamic equilibrium. One indication of large deviations from equilibrium are the relatively large abundances of high-energy isomeric forms of species. For example, both HCN and its high-energy isomer HNC are observed with relative abundances such that [HNC]/[HCN] approaches or exceeds unity in some molecular clouds. At a canonical temperature of 20 K, the equilibrium abundance ratio is expected to be many orders of magnitude less than 1. The presence of many reactive free radicals and molecular ions in interstellar gas also indicates non-equilibrium conditions.

Further evidence that interstellar molecular abundances are not controlled by thermodynamics is found by consideration of the well-known reaction

$$CO + 3H_2 \rightarrow CH_4 + H_2O.$$

At 20 K, the equilibrium constant for this process is calculated to be 10^{500} molecules^{-2} cm^6. Given a typical interstellar gas density of 10^5 particles cm^{-3}, a complete conversion of CO to CH$_4$ should occur in molecular clouds if chemical equilibrium prevailed. In contrast, CO is the second most abundant interstellar gas-phase molecule, so abundant, in fact, that it is used to map the distribution of molecular clouds in our galaxy and in external galaxies. Closely related to the high abundance of interstellar CO is the presence of polycarbon molecular species in unsaturated forms, in particular the long polyacetylene chains. These phenomena can all be explained by considering a chemical environment that is kinetically controlled, as opposed to thermodynamically controlled. The low-temperature, low-density conditions present in molecular clouds in fact favor a chemistry governed by kinetic effects.

The types of chemical reactions that can occur in interstellar clouds are limited by the physical environment in these objects. Although these regions are dense by interstellar standards, they are extremely rarified in comparison with conditions that can be obtained in terrestrial laboratories. This low density limits any chemical reaction, restricting it to a two-body process, whereas most reactions in the laboratory involve three bodies. These reactions have negligible activation energies because of the strong attraction between the positively charged ion and the neutral molecule. The energy released in the association of these two molecules drives the reaction at the low temperatures of the ISM. Processes with activation barriers generally will not occur within the lifetime of a molecular cloud (typically a million years).

One type of chemical reaction that fulfills interstellar criteria (i.e., two-body process, low activation energy barrier) are positive ion-molecule reactions of the general form

$$A^+ + B \rightarrow C^+ + D.$$

Because of the attractive force between a positive ion and a neutral species, these processes generally lack significant activation energies and have relatively fast rates, despite the fact that a third body is not participating to stabilize the products. The ion-molecule rate is also usually independent of temperature.

For ion-molecule reactions to occur, however, positive molecular ions must be present initially. The bulk volume of the dense clouds is not penetrated by starlight; thus, the energy source for generating such ions is high-energy cosmic rays. Because the bulk composition of any given molecular cloud will be molecular hydrogen and atomic helium, cosmic-ray (100-MeV)-induced ionization produces primarily H_2^+ and He^+ cations. The secondary reactions then proceed via ion-molecule processes. One important process is the reaction of H_2^+ with H_2, which occurs at the typical ion-molecule rate of $k = 2 \times 10^{-9}$ cm^{3}s^{-1} molecule^{-1}:

$$H_2^+ + H_2 \rightarrow H_3^+ + H.$$

The reaction of He^+ with H_2, in contrast, is too slow to be significant. If it were faster, it would immediately destroy the highly reactive He^+ and therefore essentially quench the formation of organic molecules.

CO, which is formed by ion-molecule reactions in molecular clouds, is present in relation to molecular hydrogen at a ratio of about 10^{-4}. Reactions with CO are also important in the ion-molecule scheme. CO reacts rapidly with H_3^+ and He^+:

$$H_3^+ + CO \rightarrow HCO^+ + H_2.$$
$$He^+ + CO \rightarrow C^+ + O + He.$$

The product of the first reaction, HCO^+, is extremely stable and is a major ion in dense molecular clouds. It has been used as a tracer of ionization and of course ion-molecule chemistry, given that, until very recently, H_3^+ had not been observed. (The detection of H_3^+ was eventually accomplished by infrared absorption spectroscopy, and the observed abundance in dense and diffuse clouds is in excellent agreement with theoretical calculations.) The second reaction is the primary basis for the very rich organic chemistry observed in interstellar clouds, because it leads to the production of C^+. In this process, the He^+ formed by the cosmic-ray ionization of He generates a C^+ in a process 10^3 times faster than the direct cosmic-ray ionization of CO. This result is a direct consequence of the lack of reactivity of He^+ with H_2.

C^+ is an important product because it can insert itself into other carbon-containing molecules to increase carbon chain length. For example, the typical synthesis for building larger organic molecules involves the reaction of hydrocarbon radicals such that:

$$C_nH_2 + C^+ \rightarrow C_{n+1}H^+ + H,$$

followed by the addition of two hydrogens, leading to $C_{n+1}H_3^+$. The neutral species is finally generated by dissociative electron recombination or by proton transfer to a suitable base. Such carbon insertion reactions most likely lead to the wide variety of carbon chains found in interstellar gas.

Ion-molecule radiative association becomes increasingly efficient with increasing molecular size. This type of process may also lead to the larger organic species. For example, methanol is thought to be created via the radiative association process

$$CH_3^+ + H_2O \rightarrow CH_3OH_2^+ + photon,$$

followed by dissociative electron recombination:

$$CH_3OH_2^+ + e^- \rightarrow CH_3OH + H.$$

Other ion-molecule reactions can create even larger compounds. Methyl formate is synthesized from

$$CH_3OH_2^+ + HCOOH \rightarrow H_2COOCH_3^+ + H_2O.$$

The neutral species is then produced again by proton transfer to a suitable base or by dissociative recombination with an electron, creating $HCOOCH_3$ and H. The main point is that, in principle, gas-phase ion-molecule reactions can create organic molecules in interstellar gas. Extensive chemical modeling of ion-molecule chemistry has been carried out and has been relatively successful in reproducing the abundances observed in interstellar space. It is not known, however, what degree of molecular complexity can be achieved through such reactions. This uncertainty remains one of the open questions for astrochemistry.

Although the bulk of the reactions in the ISM are initiated by cosmic rays or ultraviolet light, some reactions are believed to be initiated by neutral free radicals. For example, the amount of cyanoacetylene (HC_3N) present is modeled more accurately by the addition of a cyano radical ($CN^\cdot$) to acetylene than by ion-molecule reactions:

$$CN^\cdot + HC{\equiv}CH \rightarrow HC{\equiv}C{-}CN + H\cdot$$

where the symbol $\cdot$ indicates a free radical.

It is worth noting that ion-molecule reactions are believed to be a route by which significant deuterium, and some ^{13}C, enhancement occurs in organic molecules in the ISM. This fractionation effect arises from small differences in the molecular binding energies.[3] For example, the ratio of deuterated isotopomers to their normal counterparts may be enhanced by up to four orders of magnitude compared to elemental deuterium/hydrogen abundances in the ISM.[4]

Surface Reactions in the Interstellar Medium

Reactions of compounds on dust grains are another source of organic compounds in the ISM. The dust grains are produced in the circumstellar shells of red giant stars that condense from the hot material (mainly silicates) emitted from the stellar surface. Carbon stars eject carbon and partially hydrogenated carbon into the ISM by this route. Stellar ejecta from supernovas are also believed to be a source of grains. A mantle of ice consisting of water, CO, CO_2, and organics condenses on these grains at the low temperatures (~10 K) present in the dense clouds (Table 2.2).

Ultraviolet light in the ISM initiates reactions that lead to the formation of more complex structures from the simpler compounds in the mantle on the grains. One source of the ultraviolet is the radiation emitted by molecular hydrogen, following collisional excitation by electrons produced by cosmic-ray ionization.[5] The 100- to 200-nm-wavelength light has an average flux of 10^3 photons $cm^{-2}s^{-1}$ that is about 10^{-5} that of the ultraviolet flux in the diffuse ISM. Stars are a second source of ultraviolet light that is impinging on the dust in the outer regions of the dark ISM. A third source is the ultraviolet emissions from young stellar objects (newly formed stars) in the dark ISM that irradiate the dust in their vicinity. The radiation not only initiates chemical reactions but also causes the evaporation of the icy mantles from the dust grains.

The ultraviolet processing proceeds by dissociating the molecules in the mantle into free radicals. These reaction intermediates are stable at 10 K, but if the grain is warmed by absorption of additional ultraviolet photons, the radicals move around in the mantle and react with the other molecules present. If the reaction is exothermic, the

TABLE 2.2 An Inventory of Interstellar Ices Based on Infrared Spectroscopy

Species	Dark Cloud	Young Stellar Objects	
		Low Mass	High Mass
H_2O	100	100	100
NH_3	≤10	≤8	2-15
CH_4	—	<2	2
CO	25	0-60	0-25
CO_2	21	20-30	10-35
CH_3OH	<3	≤5	3-30
H_2CO	—	<2	2-6:
HCOOH	—	<1	2-6:
XCN	<1	0-2	0-6
OCS	<0.2	<0.5	0.2

NOTE: Abundances are expressed as percentages of the H_2O abundance for three categories of sight-line. A range of values generally indicates real spatial variation; where followed by a colon, it may merely reflect observational uncertainty. Values for XCN, an unidentified molecule containing C≡N bonds, are based on an assumed band strength (D.C.B. Whittet, P.A. Gerakines, J.H. Hough, and S.S. Shenoy, "Interstellar Extinction and Polarization in the Taurus Dark Clouds: The Optical Properties of Dust Near the Diffuse/Dense Cloud Interface," *Astrophysical Journal* 547(1): 872-884, 2001). A dash indicates that no data are currently available. SOURCE: After D.C.B. Whittet, *Dust in the Galactic Environment*, 2nd Edition, Series in Astronomy and Astrophysics, Institute of Physics Publishing, London, U.K., 1992.

energy released may evaporate the grain mantle, thus ejecting the compounds present into space. Mantles may also be heated and evaporated by collisions with other grains and the shock waves from supernovas.

The presence of silicates in the grains together with water, CO, and CO_2 in their mantles, with smaller amounts of methanol, formaldehyde, formic acid, and methane, has been detected by infrared spectral studies.[6,7]

The C-H stretching frequency of organics is visible in absorption bands at 3.4 μm characteristic of CH_3 and CH_2 groups. The similarities between the aliphatic C-H stretch region as seen in spectra of dust clouds in our own galaxy and as seen in the spectra of more distant galaxies suggest that the organic component of the dust in the ISM is widespread and may be an important universal residue of abiotic carbon.

The formation of molecular hydrogen from hydrogen atoms cannot occur by binary gas-phase reactions. However, hydrogen atoms have high surface mobility on the grain mantle and can readily combine on a surface to create H_2. The energy released in the process of H_2 formation is sufficient to desorb the molecule from the grain surface, even at 10 K.

The reduction of CO to formaldehyde and methanol is unlikely to occur by the action of cosmic rays because of the high activation energy required for the reactions.[8] In addition, some researchers have suggested that the amount of methanol in the grain mantles (5 to 10 percent of the water present) is much greater than that in the gas phase, suggesting that the mantle methanol was formed in the solid phase,[9,10] a claim that is in conflict with the previously proposed synthesis of methanol in the gas phase by radiative association (see above). It is likely that methanol is formed by both processes. It is possible that the reduction of CO in the mantle by hydrogen atoms is the source of the formaldehyde and methanol. The reaction with hydrogen atoms has no activation energy because hydrogen atoms can tunnel through the activation barriers on the mantle surface.

Extensive laboratory studies have shown that the ultraviolet irradiation of simulated grain mantles results in the generation of more complex organics.[11-14] Unfortunately, the laboratory studies are by necessity carried out with a high ultraviolet flux and thus are not representative of interstellar conditions. Since the precise composition of the grain mantles is not known, it is not possible to accurately extrapolate from the laboratory simulations to the amounts of these compounds in the dark ISM. For example, the higher yields of amino acids formed in the experiments of Munoz Caro et al.[15] probably reflect the use of a 10-fold lower ratio of water to the other reactants than was used in the comparable study by Bernstein et al.[16]

It is difficult to compare the extent of formation of organics in the ISM by comparison of the products formed by cosmic rays and ultraviolet. The bulk of the compounds listed in Table 2.1 were formed in gas-phase reactions driven by cosmic rays. The presence of these compounds in the ISM was determined by high-resolution radio astronomy, a technique that is very sensitive and also makes it possible to determine the structures of the compounds. Infrared spectroscopy is much less sensitive than radio astronomy and is a technique that provides information about the functional groups in the organics and not an exact structure when a mixture of compounds is present. The different characteristics of the two spectral measurements make it difficult to compare the amounts and the diversity of compounds formed.

A large number of modeling programs exist to predict gas-phase reactions and molecular abundances. The extension of the general chemical modeling programs to surface chemistry faces a number of problems and uncertainties not encountered with gas-phase binary reactions. The variables in the kinetic models are generally the densities of the reactants in the gas phase. The incorporation of surface reactions with known gas-phase reactions into a master reaction scheme presents some significant difficulties. There is a large asymmetry in the fundamental understanding of binary (gas-phase) encounters and processes on surfaces. Several problems require experimental and theoretical resolution before a quantitative model, such as discussed for the gas-phase chemistry ion-molecule chemistry, will be obtained for surface reactions. The problems include the following:

- The size and surface area of the grains as well as the chemical composition are not well characterized. It is difficult to model a surface of unknown composition.
- The exact mechanisms for reactions on surfaces are not well characterized. Also, it is difficult to find desorption processes for reaction products that are effective at low temperatures (10 K).
- The usual gas-phase reaction rate theory is not applicable to gas–surface reactions. Probability theory must be applied, and hence there are no exact solutions.

Broad Interstellar Features and the Organic Inventory

Broad emission features have been routinely observed in molecular clouds at infrared wavelengths, using spectroscopic techniques. The origin of these features, known as unidentified infrared bands, are most likely emissions from polycyclic aromatic hydrocarbons (PAHs). They occur at wavelengths that are suggestive of both the aliphatic and aromatic C-H stretching frequencies, as well as C-H deformation modes. These data indicate that organic material is present in interstellar gas that consists of large unsaturated hydrocarbons. Indeed, as already mentioned in the previous section, the spectral similarities between the aliphatic C-H stretch feature seen in interstellar dust in our galaxy and corresponding features seen in the spectra of more distant galaxies suggest that the organic component of the dust may represent an important universal residue of abiotic carbon.[17]

So-called diffuse interstellar bands—i.e., spectral features arising due to the absorption of visible light—are also observed. Hundreds of these bands have been observed, but none have been assigned to specific compounds. Currently PAHs appear to be the most likely structures absorbing the visible light, but this assignment remains to be verified.

Carbon stars are the sources proposed for the presence of PAHs in the ISM. The emission spectra of interstellar dust clouds indicate that PAHs are widespread but contribute only 5 to 10 percent of the total carbon. They have been found in interstellar dust grains, in unequilibrated chondrites, and in the martian meteorite ALH84001.

Proposed Research on Organics in the Interstellar Medium

Ion-molecule reactions are the fastest gas-phase processes known. Their properties make them prime candidates for producing interstellar molecules. Consequently, understanding these reactions is essential for evaluating the chemistry of the ISM, especially considering that this is a low-temperature environment. Reaction rates are not known for many ion-molecule, radiative association, and even certain neutral-neutral reactions that involve rather abundant interstellar carbon-bearing species. Nor are many of the branching ratios known for the products of dissociative electron recombination reactions, the main mechanism by which neutral organic species are produced. How material formed initially in dense, interstellar molecular clouds and evolved through star formation and subsequent nebular condensation is at present highly speculative. Reaction rates should be measured experimentally in the laboratory, especially at low temperatures. Also, theoretical calculations of reaction rates and reaction potential surfaces would be helpful for those processes that are too difficult to be determined by experiment, or for comparison with the experimental results. These data will enable models of interstellar chemistry to be more accurate in the calculation of abundances and in the prediction of possible new organic species. Such data will also help elucidate the major reaction pathways for the production of carbon-bearing molecules. In the laboratory, high-resolution infrared spectral measurements, including pure vibrational and rovibrational studies, of possible organic molecules will suggest other possible interstellar organics. Investigations are needed of carbon-bearing radicals and ions that might function as reaction intermediates in interstellar processes. The data obtained will enable astronomers to study additional carbon-bearing compounds in the ISM and therefore complete the inventory of organic material outside the solar system. The laboratory investigations thus should be followed up with the appropriate astronomical studies, using available telescope facilities, both ground-based and future air- and space-borne platforms such as the Stratospheric Observatory for Infrared Astronomy (SOFIA) and Herschel, as well as the Spitzer Space Telescope. Laboratory work includes both gas-phase and solid-state experiments. Astronomers additionally need to establish more complete databases for organic compounds in interstellar objects. Currently, molecular abundances are known for only a small subset of sources, and often only one such object. Hence, it is currently impossible to evaluate the diversity of organics in the ISM. Systematic observations of the key organic compounds in a statistical sample of molecular sources will be helpful in this regard.

PROTOPLANETARY DISKS

The early evolution of a young stellar object proceeds with rapid and dramatic changes.[18] Stars begin their lives in molecular clouds. As the cloud starts to fragment and collapse, a dense opaque protostellar core forms, typically a few thousand to 10,000 astronomical units (AU) across, and this core falls inward, supplying material (dust, gas, ices—including a rich array of organic molecules) to a central star. Because it is difficult to remove angular momentum from the gas during infall, the material accretes onto a rotationally supported disk surrounding the protostar. Dozens of these protoplanetary disks have been observed.

This main accretion phase is often simultaneously accompanied by prominent outflows of material (jets). When the star has accreted approximately 90 percent of its final mass, it will become a pre-main sequence star, just below the mass/temperature limit for hydrogen fusion. The disks can be up to a few hundred AU across with low densities (10^6 particles cm^{-3}) in the outer region and with densities increasing to 10^9 particles cm^{-3} near 100 AU. Temperatures remain low (~10 K) at these distances but increase close to the central protostar.

The evolution of the core and formation of the disk around the star are only broadly understood and not yet well constrained by observations. The evolution of high-mass stellar systems and low-mass systems proceeds somewhat differently. Less is known about high-mass star formation because most of the formation phase occurs while the star is embedded in an optically thick cloud of material and is therefore unobservable. One epoch in the formation of high-mass stars that has been observed is the so-called hot-core phase. In this transition phase before a hot massive young stellar object ionizes its surroundings, the object just begins to heat the surrounding neutral gases and can vaporize grain mantles.

A hot-core phase can also occur during the formation of low-mass stars like the Sun, if these stars are formed in the proximity of a massive star. The radiation from the massive star will vaporize the icy grain mantles of the small protostar and generate a hot core. The volatiles released from the hot core of a small or massive star will be subjected to the ultraviolet radiation that drives the formation of more complex organics from the volatiles released from the icy mantles.

The study of the chemical processes taking place in protoplanetary disks is limited by the angular resolution of submillimeter instrumentation, although the construction of larger submillimeter telescopes and more sensitive arrays, such as Atacama Large Millimeter Array (ALMA) in Chile, will help tremendously at these wavelengths. In addition, the Spitzer Space Telescope, SOFIA, and the James Webb Space Telescope will help in the mid and far-infrared. Numerical simulations and chemical models are able, in combination with observations, to help examine the chemistry in the disks, although there may be differences in the chemical processes for high- and low-mass objects. The chemistry and chemical processes in young stellar objects may be considered in several different regimes as shown in Table 2.3.

TABLE 2.3 The Chemistry and Chemical Processes in Young Stellar Objects

Components of the Protoplanetary Disk	Molecules Detected in Millimeter/Submillimeter Wavelength Region	Molecules in the Infrared Wavelength Region	Principal Chemical Processes Believed to Occur	Formation Stage of the Protoplanetary Disk
Dense cloud	Molecular ions, carbon chains HC_3N, CH_3OH, SO, SO_2	Simple ices H_2O, CO_2, CO, CH_3OH, HCOOH, H_2CO	Low-temperature chemistry Ion-molecule reactions	Cloud fragmentation collapse to protostar
Cold envelope around protostar	Simple species H_2CO	Ices H_2O, CO_2, CH_3OH	Low-temperature chemistry Grain surface reactions	Enters main accretion phase (class 0)
Inner warm envelope	High-excitation temperatures	High gas/solid ratios C_2H_2	Sublimation, gas-phase reactions	Protostar and disk accretion (class I)
Outflows (T-Tauri)	Ions/radicals CN, CCH, CO^+	Atomic and ionic lines CO	Shocks, sputtering, photodissociation, ionization	Protostar accretion, bipolar outflows, envelope dissipation (class II)

SOURCE: Data from E.F. Van Dishoeck and F.F.S. van der Tak, "Chemistry in Envelopes Around Massive Young Stars," pp. 97-112 in *Astrochemistry: From Molecular Clouds to Planetary Systems*, IAU Symposium 197 (Y.C. Minh and E.F. van Dishoeck, eds.), Astronomical Society of the Pacific, San Francisco, Calif., 2000.

Low-Temperature Chemical Processes in Protoplanetary Disks

In the cold regions of the precursor molecular cloud, the chemistry is dominated by ion-molecule reactions, resulting in small radicals and unsaturated molecules. As the disk warms up, molecules are released from the grains via sublimation, and this initiates gas-phase reactions, which can produce molecules such as H_2CO, C_2H_2, CH_3OH, and others.[19] These processes can lead to deuterium fractionation in the disk, so that the high deuterium/hydrogen ratios seen in comets may not necessarily imply preservation of interstellar material. Numerical simulations and chemical models will benefit greatly from high spectral and spatial observations made from new interferometric millimeter arrays (e.g., ALMA). This new observing tool will enable an understanding of potential chemical gradients in the disks in planet-forming zones. It should be noted that, once beyond the cold-core stage, the chemistry of the protostellar disk cannot be understood outside the context of a specific dynamical disk model.

High-Energy Processes in Protoplanetary Disks

Dynamic magnetic fields in young stellar objects can lead to violent reconnection phenomena (analogous to solar flares) that accelerate particles to high energies (MeV to GeV; i.e., similar to cosmic-ray energies), which can heat gases to x-ray temperatures. This process has been observationally detected from satellites with x-ray detectors and from nonthermal radio continuum radiation. In addition to heating, the x-rays cause ionization and excitation of molecules. The secondary electrons from the ionization can produce molecular ions such as H_3^+ and HeH^+. Millimeter emissions from CO, HCN, CN, and HCO^+ have been seen around young stellar objects and attributed to x-ray-induced chemistry.

The x-ray ionization dominates the ionization caused by external cosmic rays out to about 100 to 1000 AU. High-spatial-resolution spectra can distinguish the mechanism inducing the chemistry because the cosmic-ray and x-ray ionizations operate on different spatial scales in the disks. In addition, x-ray irradiation of disk dust grains, which may contain a variety of carbonaceous compounds (such as PAHs, aliphatic hydrocarbons, and so on) can result in dehydrogenation and breaking of aromatic rings. External cosmic-ray irradiation does not play much of a role in the dense inner disk, but beyond 10 AU, where the density becomes low enough that the rays are not completely attenuated, cosmic-ray irradiation produces H_3^+ and He^+ ions that can convert CO and H_2 to CO_2, CH_4, NH_3, and HCN.

Shock Waves in Protoplanetary Disks

Shock waves occur in protoplanetary disks where the outflow from the protostar collides with the surrounding cloud material, and there are also accretion shocks from the material infalling onto the disk. The accretion shocks occur where the material rains down on the disk at speeds greater than the local speed of sound. The shock waves compress and heat the gas and can therefore affect the chemistry in the disk. In high-speed shocks (producing abrupt discontinuities in the conditions in the gases), temperatures can reach 10^4 to 10^5 K, and molecules will dissociate. These can reform in the warm wake of the shock. Ices can recondense as an amorphous solid on cold grains, and this process will result in enhanced volatile trapping in the ices. Thus a mixture of unaltered and modified grains in the disk can result. In lower-speed shocks, temperatures are not as high ($\sim 10^3$ K), and endothermic reactions can produce new species not usually seen in the ambient medium. In addition, in the low-velocity shocks, ices can be removed from grain mantles as grains are driven through the medium via sputtering.

Theoretical modeling of protoplanetary disks is in its infancy, and much more research is required in order to predict the chemical reactions of the interstellar molecules when subjected to the energy sources associated with the process of star and planet formation. Laboratory research is also needed on ion-molecule chemistry, photochemistry, and reactions in shock-heated gases that model the changing conditions in solar system formation. These interdisciplinary studies may best be carried out in collaborative efforts involving planetary scientists, astronomers, and chemists.

INTERPLANETARY AND INTERSTELLAR DUST

Interplanetary dust particles (IDPs), which are formed by impacts between asteroids and by the sublimation of cometary materials, provide an opportunity to study extraterrestrial organic chemistry. Multiple flights to Earth's stratosphere (at altitudes of ~20 km via the use of U2 aircraft) have yielded a relatively large collection of IDPs ranging in size from ~5 to 50 μm. Given the small size of these particles, efficient radiative heat transfer successfully offsets the frictional heating developed during atmospheric entry, thus minimizing thermal alteration of both inorganic and organic phases and preserving what may be the most pristine extraterrestrial organic matter accessible for study on Earth. The integrated flux has been estimated at 4×10^{10} gC/yr.

IDPs are classified as anhydrous or hydrous. Anhydrous IDPs contain silicate glass and minerals such as pyroxene and olivine. A cometary origin is likely[21] and would have protected these particles from hydrothermal processing on parent bodies. Consequently, they may be the most primitive material available for study on Earth. Hydrous IDPs are dominated by clays and probably derive from asteroids.[22] Both types of IDPs contain carbon at abundances ranging up to 90 percent by weight.[23,24]

Analysis of this carbon—even determining whether it is an oxide, organic matter, or graphite—is difficult because of the small particle size. This problem has been partly circumvented by use of Fourier transform infrared (FTIR) microspectrophotometry, analytical transmission electron microscopy (TEM; with electron energy loss spectroscopy to derive chemical information), and scanning transition x-ray microscopy (STXM), the last utilizing synchrotron-based soft x-ray sources in order to examine the carbon-1s absorption edge. These techniques have shown that the organic matter in both anhydrous and hydrous IDPs is similar in that both types include aromatic and aliphatic carbon skeletons as well as ketones and carboxylic acids.[25,26] In comparison with extraterrestrial organic matter in carbonaceous chondrites, the organic matter in IDPs is, in general, less aromatic and more oxidized (predominantly as carboxyl groups). Studies employing an ion probe[27] as well as analytical TEM[28] reveal that the abundance of nitrogen in IDP organic matter is considerably greater than that observed in organic residues from carbonaceous chondrites.

One of the more intriguing aspects of organic matter in IDPs, revealed by the recent development of powerful microscopic analyses,[29-32] is the level of microscopic heterogeneity, in terms of both organic structure and isotopic abundances (e.g., H/D and $^{14}N/^{15}N$). Significant variation in aromatic, aliphatic, and carboxyl concentrations has been revealed using analytical TEM and STXM[33,34] (Note that micro-FTIR lacks the spatial resolution to reveal such spatial heterogeneity in functional group distribution.) Enormous hydrogen and nitrogen stable isotopic anomalies (^2H and ^{15}N) have also been observed. Hydrogen isotopic anomalies have been correlated with organic-rich domains.[35] Work by Keller et al.[36] concludes that aliphatic carbon is the dominant carrier of deuterium. Moreover, high-deuterium anomalies have also been correlated with organic-poor regions of IDPs, i.e., hydrated silicates.[37] Nitrogen isotopic anomalies correlate spatially with the organic phases,[38] and recent analytical TEM reveals that this nitrogen is likely an amine, possibly a substituent on aromatic moieties.

Some anhydrous IDPs have revealed localized deuterium anomalies (measured relative to hydrogen and normalized to the deuterium:hydrogen ratio in Earth's oceans) as high as 11,000‰ (1‰ is 1 part in 1,000),[39] and, in extreme cases, entire IDPs record bulk deuterium anomalies as high as ~25,000‰. Such high deuterium contents are probably derived from molecular-cloud material. The record of solar system evolution encoded by organic matter in IDPs may therefore exceed that recorded by the organic constituents of carbonaceous chondrites. Most of the advances in IDP research have occurred in the past several years. In the near future, the application of new technologies—e.g., the nanoscale secondary ion mass spectrometer (NanoSIMS)—and advances in synchrotron-based instrumentation are likely to yield further, highly significant results.

Interstellar (as opposed to interplanetary) dust grains are less well characterized. While the latter are believed to have formed in the solar system, the former formed via condensation in circumstellar regions around evolved stars, including red giants, carbon stars, asymptotic giant branch stars, novas, and supernovas. Information on the nature of interstellar grains is available from two sources, astronomical observations and laboratory studies of meteorites. The astronomical evidence is derived primarily from observations of the infrared emissions from the dust itself or studies of the dust's ability to scatter, polarize, or redden the light of background stars (the so-called interstellar extinction). Laboratory studies of meteorites have revealed nanoparticles of, for example, diamond

(300 parts per million), silicon carbide (~5 parts per million), graphite (~1 part per million), corumdum (30 parts per billion), and silicon nitride (2 parts per billion), whose peculiar isotopic compositions suggest that they formed in interstellar space and were later incorporated into the solar system during its formation.[20] Although meteoritic grains provide a window on processes occurring in the stellar environments in which they were created and the subsequent chemical evolution of the galaxy, they cannot be considered representative of all interstellar grains because of unknown selection effects.

Several characteristic features of the interstellar extinction curve—i.e., the plot of interstellar reddening as a function of wavelength—provide clues to the identity of interstellar grains. These features are as follows:

- The general shape of the extinction curve at far-ultraviolet wavelengths requires grains smaller than 0.01 µm in size.
- A very prominent broad hump in the extinction curve at ~0.22 µm is usually attributed to some form of graphitic carbon. This explanation is not, however, universally accepted.
- The general shape of the extinction curve at visible wavelengths requires grains larger than 0.1 µm in size.
- A strong absorption in the extinction curve at ~9.7 µm and another feature at 18 µm can be best explained by silicates. Most of the silicates, perhaps as much as 95 percent, are in an amorphous as opposed to a crystalline form. While the composition of the silicate remains uncertain, olivine ($MgFeSiO_4$) has been suggested as a likely candidate. Whatever the exact nature of the silicates, they almost certainly represent a considerable fraction of the mass of the interstellar dust.

In addition to the graphitic carbon and silicates, other astronomical evidence has led researchers to suggest the presence of a variety of other grain candidates, including diamonds, ultraviolet-processed hydrogenated amorphous carbon, onion-like hyper-fullerenes, glassy carbon, aliphatic hydrocarbons, polycyclic aromatic hydrocarbons, carbonates, and silicon carbide. In summary, no one particular set of chemical or physical characteristics fits all of the available evidence.

Research Opportunities

An outstanding problem in the study of IDPs involves characterization of the origins of micron-scale chemical and isotopic heterogeneity and relating this information to models of nebular formation and evolution. Does the heterogeneity, for example, reflect sources spanning a large range of heliocentric distances? Additionally, it is crucial to establish whether the organic matter of IDPs is similar to that in comets or more similar to pristine organic matter in carbonaceous chondrites. Thus, both cometary sample-return missions and extension of meteorite studies to a broader suite of carbonaceous chondrites should parallel the acceleration of the analyses of archived IDPs.

The Stardust mission successfully collected particles from Comet Wild 2 and returned them to Earth in January 2006. The chemical composition of these particles will provide insight into the importance of comets as a reservoir of organic compounds. For more information, see the section "Summary of Past, Present, and Planned Missions: Implications for Carbon Studies" in Chapter 4.

The present program for the collection of interplanetary dust is piggybacked on flights that have other scientific objectives. Consequently, the aircraft used for dust collection are seldom flown at times when Earth is passing through high-dust regions of space. If the flights are scheduled for times when Earth passes through the trails of a comet (Table 2.4), it will be possible to collect larger amounts of dust from a specific source. For example, the Leonid meteor shower in November of each year represents dusty debris from the short period comet Tempel-Tuttle. Dust collected during this meteor shower is likely to provide samples of material from this comet. The meteoritic material deposited in the atmosphere by a particular meteor shower is microscopic in size and slowly settles in the atmosphere over an extended period of time. The rate at which particles from a particular shower reach the altitude at which they will be collected by aircraft can be estimated using the Stokes-Cunningham law.[40] The settling times vary with the diameter of the particles and, thus, the average silicate sphere with a

TABLE 2.4 Principal Meteor Showers and Their Associated Comets

Name of Shower	Active Period	Location of Radiant	Source
Quadrantids	January 1-5	Bootes	96P/Machholtz and 1491 I?
April Lyrids	April 16-25	Hercules	C/Thatcher (1861 G1)
Eta Aquarids	April 19-May 28	Aquarius	1P/Halley
Arietids[a]	May 29-June 19	Aries	96P/Machholtz and 1491 I?
Zeta Perseids[a]	June 1-17	Taurus	2P/Encke[b]
Beta Taurids[a]	June 7-July 7	Taurus	—
Alpha Capricornids	July 3-August 19	Capricorn	—
S Delta Aquarids	July 15-August 28	Aquarius	96P/Machholtz and 1491 I?
Perseids	July 17-August 25	Perseus/Cassiopeia	109P/Swift-Tuttle
Kappa Cygnids	August 3-31	Draco	—
S. Taurids	September 15-November 25	Taurus	2P/Encke[b]
N. Taurids	September 15-November 25	Taurus	—
Orionids	October 2-November 7	Orion	1P/Halley
Draconids/Giacobinids	October 6-10	Draco	21P/Giacobini-Zinner
Leonids	November 14-21	Leo	55P/Tempel-Tuttle
Geminids	December 7-17	Gemini	(3200) Phaethon[c]
Ursids	December 17-267	Ursa Minor	8P/Tuttle

[a]Daytime showers.

[b]Also associated with various asteroids.

[c]Possible extinct comet or asteroid.

SOURCE: Data from R.P. Binzel, M.S. Hanner, and D.I. Steel, "Solar System Small Bodies," pp. 315-337 in *Allen's Astrophysical Quantities*, 4th edition (A.N. Cox, ed.), Springer-Verlag, New York, 2000.

diameter of 10.4 μm has a settling time of about 12 days. It will also be helpful to compare the analytical data from dust particles attributed previously to a particular comet to confirm the connection between the dust particle and the comet.

Enhancement of the stratospheric-collection program would yield a range of materials great enough to allow meaningful comparisons with putative IDPs collected from seafloor sediments and Antarctic ice sheets. Effects of alteration in those environments may be recognized and quantified, thus providing a greatly expanded time scale and more broadly representative database.

Recommendation: A program specifically designed to collect dust in the stratosphere during meteor showers should be implemented.

NOTES

1. For a comprehensive review see, for example, B.T. Draine, "Interstellar Dust Grains," *Annual Reviews of Astronomy and Astrophysics* 41: 241-289, 2003.

2. For a general review of interstellar chemistry see, for example, E. Herbst, "The Chemistry of the Interstellar Medium," *Annual Reviews of Physical Chemistry* 46: 27-53, 1995.

3. T.L. Wilson and R.T. Rood, "Abundances in the Interstellar Medium," *Annual Review of Astronomy and Astrophysics* 32: 191-226, 1994.

4. E. Herbst, "Isotopic Fractionation by Ion-Molecule Reactions," *Space Science Reviews* 106(1-4): 293-304, 2003.

5. S.S. Prasad and S.P. Tarafdar, "UV Radiation Field Inside Dense Clouds—Its Possible Existence and Chemical Implications," *Astrophysical Journal* 267: 603, 1983.

6. W.A. Schutte, A.C.A. Boogert, A.G.G.M. Tielens, D.C.B. Whittet, P.A. Gerakines, J.E. Chiar, P. Ehrenfreund, J.M. Greenberg, E.F. van Dishoeck, and Th. de Graauw, "Weak Ice Absorption Features at 7.24 and 7.41 Microns in the Spectrum of the Obscured Young Stellar Object W 33A," *Astronomy and Astrophysics* 343: 966, 1999.

7. E.L. Gibb, D.C.B. Whittet, A.C.A. Boogert, and A.G.G.M. Tielens, "Interstellar Ice: The Infrared Space Observatory Legacy," *Astrophysical Journal Supplement Series* 151: 35G, 2004.

8. T.J. Millar, E. Herbst, and S.B. Charnley, "The Formation of Oxygen-Containing Organic Molecules in the Orion Compact Ridge," *Astrophysical Journal* 369: 147, 1991.

9. A.G.G.M. Tielens and S.B. Charnley, "Circumstellar and Interstellar Synthesis of Organic Molecules," *Origins of Life and Evolution of the Biosphere* 27: 23-51, 1997.

10. R.J.A. Grim, F. Baas, T.R. Geballle, J.M. Greenberg, and W. Schutte, "Detection of Solid Methanol Toward W33A," *Astronomy and Astrophysics* 243: 473, 1997.

11. V.K. Agarwal, W. Schutte, J.M. Greenberg, J.P. Ferris, R. Briggs, S. Conner, C.P.E.M. Van De Bult, and F. Bass, "Photochemical Reactions in Interstellar Grains, Photolysis of CO, NH_3 and H_2O," *Origins of Life* 16: 21-40, 1985.

12. R. Briggs, G. Ertem, J.P. Ferris, J.M. Greenberg, P.J. McCain, C.X. Mendoza-Gomez, and W. Schutte, "Comet Halley as an Aggregate of Interstellar Dust and Further Evidence for the Photochemical Formation of Organics in the Interstellar Medium," *Origins of Life and Evolution of the Biosphere* 22: 287-307, 1992.

13. M.P. Bernstein, J.P. Dworkin, S.A. Sandford, G.W. Cooper, and L.J. Allamandola, "Racemic Amino Acids from the Ultraviolet Photolysis of Interstellar Ice Analogues," *Nature* 416: 401-403, 2002.

14. G.M. Munoz Caro, U.J. Meierhenrich, W.A Schutte, B. Barbier, A. Arcones Segovia, H. Rosenbauer, W.H.P. Thiemann, A. Brack, and J.M. Greenberg, "Amino Acids from Ultraviolet Irradiation of Interstellar Ice Analogues," *Nature* 416: 403-406, 2002.

15. G.M. Munoz Caro, U.J. Meierhenrich, W.A Schutte, B. Barbier, A. Arcones Segovia, H. Rosenbauer, W.H.P. Thiemann, A. Brack, and J.M. Greenberg, "Amino Acids from Ultraviolet Irradiation of Interstellar Ice Analogues," *Nature* 416: 403-406, 2002.

16. M.P. Bernstein, J.P. Dworkin, S.A. Sandford, G.W. Cooper, and L.J. Allamandola, "Racemic Amino Acids from the Ultraviolet Photolysis of Interstellar Ice Analogues," *Nature* 416: 401-403, 2002.

17. For a review, see Y.J. Pendleton and L.J. Allamandola, "The Organic Refractory Material in the Diffuse Interstellar Medium: Mid-infrared Spectroscopic Constraints," *Astrophysical Journal Supplement Series* 138: 75-98, 2002.

18. V. Mannings, A.P. Boss, and S.S. Russell, eds., *Protostars and Planets IV*, University of Arizona Press, Tucson, Ariz., 2002.

19. Y. Aikawa and E. Herbst, "Chemical Models of Circumstellar Disks," pp. 425-434 in *Astrochemistry: From Molecular Clouds to Planetary Systems* (Y.C. Minh and E.F. van Dishoeck, eds.), IAU Symposium 197, Astronomical Society of the Pacific, San Francisco, Calif., 2000.

20. For a recent review, see, for example, P.M. Hoppe and E. Zinner, "Presolar Dust Grains from Meteorites and Their Stellar Sources," *Journal of Geophysical Research* 105: 10371-10386, 2000.

21. D.E. Brownlee, "Cosmic Dust: Collection and Research," *Annual Review of Earth and Planetary Science* 13: 147-173, 1985.

22. L.P. Keller, K.L. Thomas, and D.S. McKay, "An Interplanetary Dust Particle with Links to CI Chondrites," *Geochimica et Cosmochimica Acta* 56: 1409-1412, 1992.

23. K.L. Thomas, L.P. Keller, G.E. Blanford, and D.S. McKay, "Quantitative Analyses of Carbon in Anhydrous and Hydrated Interplanetary Dust Particles," pp. 165-172 in *Analysis of Interplanetary Dust* (M.E. Zolensky, T.L. Wilson, F.J.M. Rietmeijer, and G.J. Flynn, eds.), AIP Conference Proceedings 310, American Institute of Physics, Woodbury, N.Y., 1994.

24. G.J. Flynn, L.P. Keller, C. Jacobsen, S. Wirick, and M.A. Miller, "Organic Carbon in Interplanetary Dust Particles," pp. 191-194 in *A New Era in Bioastronomy*, ASP Conference Series, Vol. 213, Astronomical Society of the Pacific Press, San Francisco, Calif., 2000.

25. L.P. Keller, S. Messenger, G.J. Flynn, S. Clemett, S. Wirick, and C. Jacobsen, "The Nature of Molecular Cloud Material in Interplanetary Dust," *Geochimica et Cosmochimica Acta* 68(11): 2577-2589, 2004.

26. G.J. Flynn, L.P. Keller, M. Feser, S. Wirick, and C. Jacobsen, "The Origin of Organic Matter in the Solar System: Evidence from the Interplanetary Dust Particles," *Geochimica et Cosmochimica Acta* 67(24): 4791-4806, 2003.

27. J. Aleon, F. Robert, M. Chaussidon, and B. Marty, "Nitrogen Isotopic Composition of Macromolecular Organic Matter in Interplanetary Dust Particles," *Geochimica et Cosmochimica Acta* 67: 3773-3783, 2003.

28. L.P. Keller, S. Messenger, G.J. Flynn, S. Clemett, S. Wirick, and C. Jacobsen, "The Nature of Molecular Cloud Material in Interplanetary Dust," *Geochimica et Cosmochimica Acta* 68(11): 2577-2589, 2004.

29. L.P. Keller, S. Messenger, G.J. Flynn, S. Clemett, S. Wirick, and C. Jacobsen, "The Nature of Molecular Cloud Material in Interplanetary Dust," *Geochimica et Cosmochimica Acta* 68(11): 2577-2589, 2004.

30. J. Aleon, C. Engrand, F. Robert, and M. Chaussidon, "Clues to the Origin of Interplanetary Dust Particles from the Isotopic Study of Their Hydrogen Bearing Phases," *Geochimica et Cosmochimica Acta* 65: 4399-4412, 2001.

31. J. Aleon, F. Robert, M. Chaussidon, and B. Marty, "Nitrogen Isotopic Composition of Macromolecular Organic Matter in Interplanetary Dust Particles," *Geochimica et Cosmochimica Acta* 67: 3773-3783, 2003.

32. G.J. Flynn, L.P. Keller, M. Feser, S. Wirick, and C. Jacobsen, "The Origin of Organic Matter in the Solar System: Evidence from the Interplanetary Dust Particles," p. 275 in *Bioastronomy 2002: Life Among the Stars*, IAU Symposium 213 (R. Norris and F. Stootman, eds.), Astronomical Society of the Pacific, San Francisco, Calif., 2003.

33. L.P. Keller, S. Messenger, G.J. Flynn, S. Clemett, S. Wirick, and C. Jacobsen, "The Nature of Molecular Cloud Material in Interplanetary Dust," *Geochimica et Cosmochimica Acta* 68(11): 2577-2589, 2004.

34. G.J. Flynn, L.P. Keller, M. Feser, S. Wirick, and C. Jacobsen, "The Origin of Organic Matter in the Solar System: Evidence from the Interplanetary Dust Particles," *Geochimica et Cosmochimica Acta* 67(24): 4791-4806, 2003.

35. J. Aleon, C. Engrand, F. Robert, and M. Chaussidon, "Clues to the Origin of Interplanetary Dust Particles from the Isotopic Study of their Hydrogen Bearing Phases," *Geochimica et Cosmochimica Acta* 65: 4399-4412, 2001.

36. L.P. Keller, S. Messenger, G.J. Flynn, S. Clemett, S. Wirick, and C. Jacobsen, "The Nature of Molecular Cloud Material in Interplanetary Dust," *Geochimica et Cosmochimica Acta* 68(11): 2577-2589, 2004.

37. L. Nittler, Carnegie Institution of Washington, personal communication, 2003.

38. J. Aleon, F. Robert, M. Chaussidon, and B. Marty, "Nitrogen Isotopic Composition of Macromolecular Organic Matter in Interplanetary Dust Particles," *Geochimica et Cosmochimica Acta* 67: 3773-3783, 2003.

39. See, for example, S. Messenger, "Identification of Molecular-cloud Material in Interplanetary Dust Particles," *Nature* 404: 968-971, 2000.

40. See, for example, F.J.M. Rietmeijer and P. Jenniskens, "Recognizing Leonid Meteoroids Among the Collected Stratospheric Dust," *Earth, Moon, and Planets* 82-83: 505-524, 1998.

3

Meteorites

THE ORIGIN OF METEORITES

Meteorites are products of collisions that occur within the asteroid belt. As such, they sample remnants of accretionary products of the early solar system. It has also been proposed that a few may be the remains of cometary nuclei, i.e., derived from outside the asteroid belt. A small number of meteorites are recognized to be samples of debris from the surfaces of the Moon and Mars, ejected by impacts.

Meteorites are divided into three main groups: irons, stony irons, and stony meteorites. The irons contain at least 90 percent metal, predominantly iron and nickel. The stony irons are further divided into pallasites, composed of crystalline minerals embedded in a metallic matrix, and mesosiderites, which contain a more fine-grained and intimate mixture of minerals and metal. The stony meteorites consist predominantly of silicates and are subclassified as either chondrites (materials that have never experienced igneous processing) or achondrites (lavas, cumulates, or residues from partial melting). Many stony meteorites contain spherules of rock ranging in size from a few millimeters to one centimeter across. These so-called chondrules are apparently the crystallization products of silicate melt droplets, whose melting history remains a subject of intense investigation. Indeed, the presence or absence of chondrules was once used as the defining characteristic of chondrites and achondrites, respectively. The chondrites are further subdivided according to their composition:

- Ordinary chondrites,
- Enstatite (magnesium silicate-containing) chondrites, and
- Carbonaceous chondrites.

All three groups of chondritic meteorites contain carbon that has been variously modified by parent body processes. Note that this carbon includes both inorganic and organic phases. It was the presence of organic matter identified in the carbonaceous chondrites that captured the interest of chemists in the 1800s.[1,2] Notwithstanding this interest, analytical techniques capable of describing the organic constituents of these meteorites were not available until the 1950s, when a rising interest in space science coupled with advances in analytical chemistry led to a rejuvenation of efforts to characterize the organic matter in carbonaceous chondrites.

CARBONACEOUS CHONDRITES: A RECORD OF THE ORGANIC CHEMICAL EVOLUTION OF THE EARLY SOLAR SYSTEM

Studies of carbonaceous chondrites have provided evidence that interstellar organic matter has survived processes associated with the formation of the solar system and has been incorporated into grains and larger objects. Within the asteroidal parent bodies, the presumably simpler interstellar organic compounds were transformed into more complex organic compounds by aqueous and thermal processing.[3-6] A remarkably broad range of organic compounds has been identified in carbonaceous chondrites.[7,8] This organic ensemble includes amino acids, purines, and pyrimidines. The presence of these particular compounds demonstrates that the solar system provided at least one environment in which recognizable biomolecules were synthesized abiotically. Furthermore, the presence of such compounds in meteorites indicates that impacts by meteorites, comets, and dust must have delivered potentially biologically useful organic compounds to early Earth and the other terrestrial planets. Carbonaceous chondrites thus currently provide a potentially rich and accessible source of information regarding the organic compounds naturally present during prebiotic evolution of the terrestrial planets as well as other solid bodies such as the moons of Jupiter and Saturn (e.g., Europa, Enceladus, or Titan).

Murchison Meteorite: A Timely Gift from the Solar System

In September 1969, just as many chemists were beginning the search for organic compounds in lunar samples, a new carbonaceous chondrite fell near Murchison, Victoria, Australia. Rich in volatiles, it contains more than 10 percent water and about 2.2 percent carbon by weight. Initial research was directed toward determining the constituents of the aqueous and solvent-soluble fractions of this meteorite. It was soon determined that Murchison contained a spectacularly complex suite of small molecules (Table 3.1). For example, to date, more than 70 different amino acids have been identified in this meteorite. The distribution of amino acids is similar to that produced in certain abiotic syntheses and, where mirror-image structures (i.e., stereoisomers) are anticipated, both chiral forms are found to be nearly equal in abundance. This balance is one signature of abiotic synthesis. Moreover, the amino acids extracted from the Murchison meteorite contain distinctly higher levels of ^{13}C than do any terrestrial amino acids. Thus, even though many of the soluble compounds in Murchison are recognizable as biochemically significant, e.g., the amino acids, their stereochemical and isotopic characteristics clearly identify them as both extraterrestrial and nonbiological.[9]

The classes of soluble meteoritic organic compounds that have familiar biochemical counterparts include the amino acids, fatty acids, purines, pyrimidines, and sugars.[10,11] Additional soluble constituents include alcohols, aldehydes, amides, amines, mono- and dicarboxylic acids, aliphatic and aromatic hydrocarbons, heterocyclic aromatics, hydroxy acids, ketones, phosphonic and sulfonic acids, sulfides and ethers.[12-14] Concentrations of the major representatives of these classes vary widely from less than 10 parts per million (amines) to tens of parts per million (amino acids) to hundreds of parts per million (carboxylic acids).[15]

Cooper et al.[16] identified a complex suite of sugars and sugar derivatives present in the soluble fraction of Murchison at concentrations comparable to that of the amino acids (~60 ppm) in the Murchison and Murray meteorites. Chromatographic analyses of virtually all classes of acyclic compounds reveal complex molecular assemblages containing homologous series of compounds up to C_{12} in some cases (carboxylic acids). The relative abundances of the heavy stable isotopes of carbon and hydrogen (^{13}C and ^{2}H) in these compounds differ significantly from those of terrestrial materials and provide decisive evidence that these materials are not terrestrial contaminants.

Distinctive patterns of structural variation are seen in the Murchison organics but they differ from those found in living systems.[17] As a class, the amino acids exemplify this contrast. Specifically:

- The abundances of amino acids decrease with increasing carbon number,[18]
- The abundances of branched-chain isomers far exceed those of the straight-chain isomers, and
- Structural diversity dominates at the lower carbon numbers (e.g., acyclic, monoamino acids).

TABLE 3.1 Distribution of Carbon in the Murchison Meteorite

Form	Amount[a]
Total carbon	2.12%
Interstellar grains	
Diamond	400 ppm
Silicon carbide	7 ppm
Graphite	<2 ppm
Carbonates	2-10% total C
Macromolecular material	70-80% total C
Small organics	
Aliphatic hydrocarbons	••
Aromatic hydrocarbons	••
Polar hydrocarbons	•••
Volatile hydrocarbons	•
Aldehydes and ketones	••
Alcohols	••
Amines	•
Monocarboxylic acids	•••
Dicarboxylic acids	••
Sulfonic acids	•••
Phosphonic acids	•
N-heterocycles	•
Purines and pyrimidines	•
Carboxamides	••
Hydroxy acids	••
Amino acids	••
Sugars and sugar derivatives	••

[a] ••• >100 ppm; •• >10 ppm; • >1 ppm.
SOURCE: Modified after J.R. Cronin, "Clues from the Origin of the Solar System: Meteorites," pp. 119-146 in *The Molecular Origins of Life: Assembling Pieces of the Puzzle*, A. Brack A. (ed.), Cambridge University Press, Cambridge, U.K., 1998.

Overall, approximately 70 amino acids have been identified to date among the 159 possible C_2 to C_7 isomers.[19,20] Living systems use a highly restricted number of amino acid isomers to fulfill requirements for protein structure and function. At present, life is known to encode only 20 protein α-amino acids, all of which have at least one hydrogen atom attached to the α carbon. Only eight out of 20 of these terrestrial, biological amino acids have been identified in meteorite extracts.

Mechanism of Formation of Organic Compounds in Carbonaceous Chondrites

The observed patterns of variation in molecular structure and abundance in carbonaceous chondrites suggest initial synthesis routes involving the formation and random recombination of small, free radicals.[21,22] Such reactions tend to produce the greatest variation in possible structural isomers at lower carbon numbers. It is generally accepted that ion-molecule and related reactions within interstellar clouds produce mixtures of nitriles and other highly reactive organic compounds (e.g., polyacetylenes). These compounds, when exposed to liquid water in the meteorite parent body, will hydrolyze to form many compounds superficially similar to those identified in the

TABLE 3.2 Soluble Organic Compounds in the Tagish Lake and Murchison Meteorites

Class	Tagish Lake		Murchison	
	Concentration (ppm)[a]	Compounds Identified[b]	Concentration (ppm)	Compounds Identified
Aliphatic hydrocarbons	5	12	>35	140
Aromatic hydrocarbons	≥1	13	15-28	87
Dicarboxylic acids	17.5	18	>30	17
Carboxylic acids	40.0	7	>300	20
Amino acids	<0.1	4	60	74
Hydroxy acids	b.d	0	15	7
Pyridine carboxylic acids	7.5	7	>7	7
Sulfonic acids	≥20	1	67	4
Nitrogen heterocycles	n.d.	n.d.	7	31
Amines	<0.1	3	8	10
Amides	<0.1	1	n.d.	4
Dicarboximides	5.5	4	>50	3

NOTE: n.d., not determined.
[a]Concentrations are based on chromatographic peak intensities and include compounds identifed by reference standards and mass spectra. Variability was not estimated at this time, as measurements were obtained by analyses of one meteorite stone.
[b]Compounds identifed with reference standards.
SOURCE: Modified from S. Pizzarello, Y. Huang, L. Becker, R.J. Poreda, R.A. Nieman, G. Cooper, and M. Williams, "The Organic Content of the Tagish Lake Meteorite," *Science* 293: 2236-2239, 2001, Table 1.

Murchison extracts. For example, nitriles will hydrolyze to yield carboxylic acids. Amino acids may have formed directly from a series of reactions referred to as the Strecker synthesis. In this reaction cyanohydrins and aminonitriles form from condensation of HCN and aldehydes or ketones. Subsequent hydrolysis yields α-amino and α-hydroxy acids. Amino acids substituted at positions more than one carbon atom away from the carboxyl group (common in the Murchison organic solubles) indicate alternative synthetic pathways.

The cosmochemical community has also benefited from another large meteorite fall, the Tagish Lake carbonaceous chondrite recovered in January 2000 on the border between British Columbia and the Yukon Territory, Canada. This meteorite fortuitously fell in the winter on snow and ice and was recovered frozen. Obtained in this frozen state, the Tagish Lake meteorite holds the promise of revealing considerably more about the inventory of small volatile compounds that may have been lost from Murchison and other falls. Analyses of a soluble fraction of the Tagish Lake stone reveal a suite of soluble organic compounds including mono- and dicarboxylic acids, dicarboximines, pyridine carboxylic acids, sulfonic acids, and aliphatic and aromatic hydrocarbons (Table 3.2). However, in contrast to carbonaceous chondrites such as Murchison, Orgueil, and Ivuna,[23,24] virtually no amino acids are detected in extractions obtained from the Tagish Lake meteorite when it was treated with water or organic solvents. One investigator has proposed that the parent body of Tagish Lake did not support formation of amino acids via the Strecker-cyanohydrin reaction due to a lack of liquid water (key to the hydrolysis step). An alternative postulate is that the parent body of Tagish Lake was formed closer to Jupiter where the composition of starting materials was different from that of other primitive bodies. These suggestions led to the speculation that Tagish Lake may represent cometary rather than asteroidal material, employing the same arguments for the designation of anhydrous versus hydrous IDPs. Clearly, considerably more analytical work is required before the sources and history of these materials can be outlined in detail.

Nitrogen Heterocycles in Carbonaceous Chondrites

The essential biological role of nucleobases, e.g., purines and pyrimidines, as base-pairing complements in RNA and DNA has led to interest in the isolation and characterization of such compounds and related derivatives in meteorites.[25-31] Although cosmochemical studies of meteorites, in particular Murchison, have provided clues about the mechanism of formation of amino acids via the Strecker synthesis,[32] the mechanisms of formation for purines, pyrimidines, and other nitrogen heterocycles remain obscure.

Stoks and Schwartz[33] noted that alkyl pyridines can be produced from aldehydes and ammonia by a reaction that is catalyzed under conditions that support Fischer-Tropsch (FT) type reactions. A similar mechanism has been postulated to be responsible for the formation of organic molecules in a cooling gaseous solar nebula, where dust grains may have provided catalytic surfaces.[34] Many of the Murchison N-heterocycles may have been produced by HCN polymerization followed by hydrolysis.[35] Notably, HCN is a critical intermediate in Miller-Urey (MU) type reactions.[36] It is present in the gas phase of the ISM and is thought to have condensed into the icy mantle of dust grains. However, key issues remain. Kung et al.[37] have shown that neither FT nor MU systems could yield N-isotopic ranges comparable to those observed in meteoritic organic matter.

The N-heterocycles identified in Murchison include:

- Purines (xanthine, hypoxanthine, guanine, and adenine),
- Pyrimidines (uracil),
- Quinolines/isoquinolines, and
- Alkyl pyridines.

The purines and pyrimidines (1.3 ppm, total) identified so far are restricted to biologically utilized entities, whereas the quinolines and pyridines are structurally diverse; i.e., they include significant quantities of isomeric, alkyl derivatives, most of which are not similar to biosynthetic products.[38]

Finally, carboxylated pyridines have been detected in hot-water extracts of the Tagish Lake carbonaceous chondrite.[39] These include isomers of nicotinic acid and 12 methyl- and dimethyl-substituted homologs. Isotopic analyses of these N-heterocycles reveal enrichment in D, ^{13}C, and ^{15}N at levels outside the terrestrial range.[40] Similar enrichments in heavy isotopes are signatures of low-temperature reactions in interstellar clouds and support an extraterrestrial source for these molecules.

Macromolecular Organic Matter in Carbonaceous Chondritic Meteorites

The majority of organic matter contained within carbonaceous chondrites is in the form of nonextractable, presumably macromolecular carbon. The meteoritic insoluble organic matter (IOM) is usually isolated as the insoluble residue that remains after solvent extractions and dissolution of minerals by treatment with HF and HCl. There has been an unfortunate tendency to refer to this insoluble organic matter as "kerogen-like."[41] However, the term "kerogen" refers to insoluble organic matter derived from biomacromolecular precursors found in ancient terrestrial sediments; kerogen therefore represents a mixture of resistant biopolymers and condensation products resulting from reactions between lipids, proteins, and carbohydrates. As there clearly is no biological connection between meteoritic IOM and terrestrial kerogen, the term "kerogen" should not be applied to the IOM of carbonaceous chondrites.

The structure of the macromolecular component in carbonaceous chondrites has proven to be difficult to ascertain. Based on early pyrolytic (thermal degradation) studies, a calculated elemental formula of $C_{100}H_{48}N_{1.8}O_{12}S_2$ has been proposed for the IOM in the Murchison meteorite.[42] Given its insoluble nature, analysis of this material is restricted either to nondestructive molecular spectroscopic methods such as FTIR and solid-state nuclear magnetic resonance (NMR) spectroscopy or to destructive methods such as pyrolysis gas chromatography–mass spectrometry and chemical degradation (e.g., oxidation with CuO, $KMnO_4$, or RuO_4). In general, the destructive analyses of IOM yield dominantly 1- and 2-ring aromatic products, generally highly alkylated, and commonly oxygen-substituted. The significant abundances of alkylated phenols in both pyrolysis and chemical degradation products has led to the conclusion that alkyl-aryl ether linkages in the IOM macromolecule are common.[43] Heteroatoms, N and S, are accounted for mostly by alkyl pyridines, benzonitrile, and alkyl thiophenes and benzothiophenes.

Hydrous pyrolysis (i.e., heating in water at high temperatures and pressures) has been used to degrade and solubilize the IOM in a number of carbonaceous chondrites.[44] In addition to aromatic hydrocarbons, alkyl-phenols are observed in significant abundance, lending support to previous conclusions that alkyl-aromatic ethers are predominant covalent linkages in the IOM macromolecule. The most striking observation based on hydrous pyrolysis experiments is the abundance of small aromatic molecules (i.e., one- and two-ring compounds) over larger polycyclic aromatics, e.g., coronene (seven rings) and larger. The preponderance of evidence based on the degradative studies indicates that the majority of molecular units that constitute the IOM macromolecule are small, functionalized aromatic moieties.

The first, nondestructive, solid-state NMR analyses of meteoritic IOM were performed by Cronin et al.,[45] wherein IOM isolated from Orgueil, Murchison, and Allende were analyzed using ^{13}C NMR spectroscopy. This early analysis led to the conclusion that the fraction of aromatic carbon in IOM was on the order of 40 to 50 percent of the total, a value that Cronin believed was too low. More recent analyses by Gardinier et al.[46] and Cody et al.[47] supported Cronin et al.'s earlier concern and revealed that in the case of Murchison, the aromatic content is ~63 percent of the total carbon.

Solid-state ^{13}C NMR analysis by Pizzarello et al.[48] of the Tagish Lake meteorite suggested that virtually 100 percent of the carbon in this meteorite's IOM is aromatic. However, subsequent analysis of the Tagish IOM (a different sample than that analyzed by Pizzarello) by Cody et al.[49] showed a yield of aromatic carbon of ~83 percent. Notwithstanding this variation, it is now universally accepted that the structure of meteoritic IOM is highly complex. Furthermore, the relative contributions of different functional groups vary enormously across meteorite groups.[50]

To date, the IOM has been extensively characterized only for Murchison,[51] wherein eight independent solid-state NMR experiments were used to provide a self-consistent assessment of chemical characteristics. First, these analyses showed that although the fraction of aromatic carbon is high (~60 percent), the average size of aromatic moieties is not large. Significant quantities (>1 to 5 percent) of graphite and/or carbon-rich domains (aside from nanodiamond that is readily observed using NMR) are absent. Second, the average fraction of aromatic carbon bonded to hydrogen is low (~30 percent for Murchison), indicating that substitution and cross-linking of the aromatic moieties are common. Third, the aliphatic carbon is highly branched [i.e., there is a significant fraction of methine (CH) carbon as opposed to methylene (CH_2)]. The majority of aliphatic carbon may be methyl-substituted alicyclic moieties. Finally, the oxygen content as determined from NMR is very high and, given the constraints of the elemental analyses, requires that most of the oxygen occurs in linkages such as ethers or esters rather than phenols or carboxyl groups.

Finally, traces of various inorganic carbon species including diamond, graphite, fullerene, and silicon carbide are present in the IOM. Within these inorganic carbon phases, many individual constituents (e.g., nanodiamond grains and SiC grains) carry large isotopic anomalies that point to specific stellar origins such as supernovas for the nanodiamonds or carbon stars and novas for SiC, fullerene, and graphite.[52-54]

Isotopic Characteristics of Insoluble Organic Matter

The bulk isotopic compositions of IOM fractions of numerous carbonaceous chondrites are well established. In general, the carbon isotopic abundance does not vary significantly in comparisons of IOM derived from various carbonaceous groups; where bulk carbon isotopic values fall within the terrestrial range of $\delta^{13}C$ at ~ −17‰. The stable isotopic composition of nitrogen, however, varies more substantially, with values ranging from those that appear clearly extraterrestrial ($\delta^{15}N$ of 200‰ to 300‰) down to values indistinguishable from terrestrial ($\delta^{15}N$ ~ −5‰). It has been proposed that the range in nitrogen isotopic abundance reflects variation in parent body processing, where the high abundances of ^{15}N correspond to more pristine (less altered) IOM.[55] Abundances of deuterium almost always deviate significantly from terrestrial levels (e.g., ~ +600‰ up to ~ +3,500‰). These suggest a chemical connection between insoluble organic matter in meteorites and ion-molecule reactions in the interstellar medium.

The carbon isotopic compositions of discrete molecules derived from degradative chemistry, e.g., hydrous-pyrolysis,[56] have also been examined. Significant variations have been observed. The abundance of ^{13}C in ethylbenzene, for example, varies by more than 20‰ between the Orgueil, Cold Bokkeveld, and Murchison carbonaceous chondrites.[57] The origin of this intermeteorite variability is not known and certainly requires further investigation.

Finally, step-combustion with simultaneous isotope-ratio monitoring also indicates nonhomogeneous isotopic distributions (e.g., C, N, and H) in the bulk insoluble organic matter of carbonaceous chondrites. The pattern of release of ^{13}C and ^{15}N is complex and varies substantially in intermeteorite comparisons. The observed variation has been interpreted as resulting from the extent of parent body processing.[58]

Chirality: Enantiomeric Excesses Observed in Stereoisomers from Carbonaceous Chondrites

A fundamental characteristic of life is the homochirality of most of its building blocks. Various theories have been proposed to explain the origin of this property. Most require a chemical-amplification scheme but differ in the origin of the initial imbalance. Experimental investigations into the abiotic syntheses of organic compounds do not produce chiral products.[59,60] Moreover, recent evaluations of abiotic mechanisms proposed for the origin of chiral molecules on the primitive Earth have concluded that such processes are not likely to occur naturally.[61]

Enantiomeric excesses have been observed among amino acids extracted from meteoritic matter[62-65] and have been sought in micrometeorites[66] and comets (i.e., the Cometary Sampling and Composition (COSAC) experiment aboard the European Space Agency mission Rosetta[67]). In 1997, Cronin and Pizzarello[68] reported modest L-enantiomeric excesses of 2 to 9 percent in some amino acids in the Murchison meteorite. They avoided the problems of contamination by making measurements on 2-amino-2,3-dimethylpentanoic acid, α-methyl norvaline, and isovaline. All three of these compounds are α-methyl substituted; the first two have no known biological counterparts and the third has a highly restricted distribution in fungal antibiotics. The α-methyl substituents are significant in that they block the known racemization of chiral amino acids. This behavior suggests that the enantiomeric excess may have occurred when these amino acids formed. Bailey and colleagues suggested that the observed enantiomeric excesses could have been induced by circularly polarized light scattered from dust in regions of high-mass-star formation.[69] These sources occur more widely than do the supernova remnants or pulsars that were first proposed by Rubenstein et al. as sources of circularly polarized synchrotron radiation.[70] However, both theories do not take into account that amino acids are very fragile compounds, which are easily destroyed by particle radiation and even by low-energy ultraviolet photons.[71]

ORGANIC CARBON IN UNEQUILIBRATED ORDINARY CHONDRITES AND ENSTATITE CHONDRITES

Most analyses of organic material pertain to Murchison and a few other carbonaceous chondrites. The unequilibrated ordinary chondrites (UOCs) are generally recognized as the most primitive ordinary chondrites. They also contain organic carbon (in excess of that contained in Murchison[72] when normalized to matrix content, although considerably less on a total meteorite basis). Historically, these interesting stones have received considerably less attention than have the carbonaceous chondrites. This disparity is due at least in part to the fortuitous fall of the large and organic-rich Murchison chondrite in 1969 and in part due to the fear of terrestrial contamination that leads most researchers to restrict their analytical studies to this most recent fall. Claims that biogenic compounds are present in UOCs span more than a century. Kaplan et al. reported the detection of amino acids and sugars in UOCs,[73] although this observation has been generally discounted as due to terrestrial contamination. Alkanes were detected in one UOC;[74] however, an analysis using compound-specific isotopic analysis of the Bishunpur UOC clearly indicates a terrestrial source of contamination.[75] Clemett et al.[76] and Kovalenko et al.[77] reported polycyclic aromatic hydrocarbons in a UOC using two-stage, laser-desorption, resonant-ionization, time-of-flight mass spectrometry. Later, Sephton et al.[78] detected toluene and dimethyl ethyl naphthalene in a hydrous pyrolysate obtained from the Bishunpur UOC; the fact that these compounds were only detected upon destruction of the macromolecular matrix supports indigeneity as opposed to terrestrial contamination. Notwithstanding these later data, the organic matter contained within UOCs remains largely unexplored.

The enstatite chondrites also contain relatively large amounts of carbon. These meteorites, however, may have been subjected to metamorphic temperatures high enough to convert all of the organic carbon to graphitic or graphite-like phases. Notably, the high content of ^{15}N suggests that the carbon in the enstatite chondrites was derived from an organic precursor. But there is no a priori reason to believe that E3 chondrites experienced any higher degree of metamorphism than did H, L, or LL3 chondrites. What is clear, however, is that enstatite chondrites have a much lower percentage of matrix and that the dominant chondrules did experience igneous temperatures during their history.

MARTIAN METEORITES

Meteorites designated by the abbreviation SNC (Shergottite-Nakhlite-Chassignite) are generally accepted as martian in origin. This assignment is based on the isotopic compositions and relative abundances of argon and other noble gases within the SNC meteorites.[79]

To date, all that is known about the organic matter on Mars comes from the examination of martian meteorites. Studies of examples of the 35 or so cataloged martian meteorites[80] have revealed much about their organic constituents. For example, stepped-combustion studies of the Shergotty meteorite and other known shergottites indicate that they contain complex organic molecules and components of both low and high thermal stability.[81] Similar studies of the shergottite known as EET A79001 revealed unexpectedly high concentrations of organic materials. However, because the abundances of the carbon isotopes are identical to that of terrestrial biogenic compounds, contamination cannot be excluded.[82] Complex organic materials with a high molecular weight also turned up in a more recent analysis of both EET A79001 and the Nakhla meteorite. On pyrolysis, the major components of this complex organic matter found in both meteorites are aromatic and alkylaromatic hydrocarbons, phenol, and benzonitrile. Analysis of individual molecules in the Nakhla pyrolysate showed that they were similar to materials found in carbonaceous chondrites. Nevertheless, terrestrial contamination still cannot be ruled out entirely.[83]

Perhaps the most famous martian meteorite is ALH 84001, which was reported to contain both "nanofossiles" and polycyclic aromatic hydrocarbons.[84] The announcement of these results has led to a reassessment of how to search for life on Mars and elsewhere. Much of the controversy surrounding ALH 84001 centers on stable carbon isotope studies of the carbonate and the associated organic matter. Thus, subsequent studies of the ALH 84001 organic matter have focused on determining the $\delta^{13}C$ values for specific organic compounds isolated from various mineral phases.

Two independent investigations of the organics in ALH 84001, stepped combustion experiments to measure the $^{13}C/^{12}C$ compositions for the organic matter, indicated that a small portion (~50 ppm out of 250 ppm) of this material yielded a $\delta^{13}C$ value of −15‰.[85,86] A $\delta^{13}C$ value of −15‰ would be unusual for indigenous martian organics based on $^{13}C/^{12}C$ measurements of trapped gases in some martian meteorites that revealed two distinct carbon reservoirs on Mars: an isotopically heavy component (atmosphere) enriched in $\delta^{13}C$ (+36‰) and a high-temperature igneous (i.e., mantle) component ($\delta^{13}C$ −20 to −30‰). On the other hand, a $\delta^{13}C$ value of −15‰ is consistent with the isotopic composition of the IOM component in carbonaceous chondrites. Thus, it is possible that some portion of the organic matter in ALH 84001 may be derived from chondritic or cometary debris exogenously delivered to the surface of Mars, and subsequently transported to Earth via the SNC carrier.

An important question, therefore, remains as to whether any of the organic matter detected in ALH 84001 is indigenous, having formed as a result of biogenic or abiogenic processes on the surface of Mars. Recent measurements of the oxygen isotopes in ALH 84001 carbonate-rich material revealed a $\delta^{17}O$ anomaly that can only be explained by a thin atmosphere and ozone production leading to a highly oxidized surface at the time that the ALH 84001 carbonates formed. The presence of highly oxidizing surface conditions was also used to explain the lack of organic compounds detected during the Viking missions. These new findings suggest that similar conditions may have existed early on in the development of the martian atmosphere. If these conditions did, in fact, occur, then the organic matter found in ALH 84001 may not have formed on the surface of the planet because the environment would have been unfavorable for abiotic synthesis and subsequent accumulation of organic compounds. This suggestion does not discount the possibility, however, that the organic matter was formed abiologically deeper within the crust and that the ^{17}O anomalous carbonate-bearing fluids did not percolate down from the surface; clearly, considerably more work needs to be done to clarify the provenance of the organic matter detected in ALH 84001 and perhaps other martian meteorites.

ORGANIC MATTER IN METEORITES: RECOMMENDATIONS

Carbonaceous meteorites are an important source of abiotic, extraterrestrial carbon that is delivered to Earth at no cost. Together with the unequilibrated ordinary chondrites, a few martian meteorites, and fragments of crust from the earliest Earth, they represent immediately available samples of great relevance to studies of organic material in the solar system. The ancient terrestrial materials have been subjected to extensive post-depositional alteration and, as far as can be discerned, already bore the imprint of biogenicity at the time of their burial. New analyses of carbonaceous chondrites would benefit from modern analytical methods (e.g., compound-specific isotopic analysis) that allow the separation of signals from contamination by terrestrial organics and signals from indigenous extraterrestrial organic matter, thus overcoming a problem that severely hindered analyses throughout the 1960s and 1970s. For example, the Murchison meteorite has for more than 30 years been the subject of numerous detailed analyses of the organic and inorganic compounds present in it. The continuing analysis of the Murchison meteorite has revealed the presence of a diverse array of different classes of organic compounds as well as distinctive isotopic compositions (see Table 3.1). A more sensitive and detailed analysis of the other carbonaceous chondrites is a cost-effective step that would be of great value in enhancing understanding of the formation of these organic materials and, therefore, yielding new information about organic-chemical processes in the early solar system. The results would provide reference points for comparison with the organics in samples returned by spacecraft missions to other bodies in the solar system.

The detection of any organic material in SNC meteorites is of considerable interest. Such studies must address issues of indigeneity before focusing on biogenicity and martian history. For broadly significant studies of organic material in the cosmos, the return to Earth of carbonaceous cometary or asteroidal material is an objective of the highest order.

How should plans be developed and proposals solicited for the curation and coordinated, intensive investigation of the composition of organic materials in carbonaceous chondrites, unequilibrated ordinary chondrites, and SNC meteorites? While the available samples are not improving on the shelf, significant developments in analytical technology have occurred in the past few decades, and so the time is ripe for investment of significant portions of the available stocks in a new round of analyses. Analyses should focus on the following:

- The location and relative abundances of the organic molecules within the mineral matrices and on mineral surfaces;
- The structural composition of all organic phases including, to the greatest extent possible, any macromolecular material;
- The isotopic compositions of all molecules and other definable subfractions; and
- The nature of contaminants and the mechanisms by which samples can become contaminated, both pre- and post-collection.

Very specifically, these investigations would require the coordination of analyses in multiple laboratories, the development of new procedures, and the upgrading of existing facilities. The effort envisioned would be on a scale completely different from that of prior analyses of carbonaceous chondrites. To provide comparability and to bring the best techniques to bear on each object, samples should be shared extensively between laboratories.

The planning of analyses, allocation of and/or access to samples, and, possibly, provision of funding could be managed by a committee of appropriately selected experts. Two extreme models of how such a committee might function are offered:

- At the centralized, highly focused extreme is the approach used for the coordinated and very intensive investigations of the Apollo lunar samples conducted between 1969 and 1985. The role of such a committee would, ideally, extend beyond the planning of analyses to include allocation and, perhaps, the provision of funding.
- At the decentralized, more informal extreme is a community-based group modeled on the Mars Exploration Program Analysis Group (MEPAG) that has successfully provided science input for the planning and

prioritization of NASA's Mars exploration activities.[87] The role of such a group is likely to be limited to that of a forum for the planning of analyses. The existing Curation and Analysis Planning Team for Extraterrestrial Materials (CAPTEM) might provide the nucleus around which an inclusive, community-based forum might be accreted.

Although the former model is likely to be the most direct approach for obtaining information about organic materials from elsewhere in the solar system, it is also the most likely to run afoul of the realities of the operation of the curatorial community. In the case of lunar samples, all materials—or most, if the samples from the former Soviet Union's three lunar sample-return missions are included—were under the control of a single organization, NASA. That is far from the case where meteorites are concerned.

The world's meteorite collections are, for all practical purposes, managed by the U.S., Japanese, and European (EUROMET) Antarctic meteorite programs; a handful of large museums—principally those in Washington, Chicago, New York, London, Paris, and Vienna—and private dealers. It is far from clear if any of these groups would cede control of their collections or any portion of their collections for distribution by a group interested in only one facet of the meteorite (e.g., organics, hydrous alteration, chondrule formation, and so on).

While it is true that all of these organizations welcome research requests and have distributed material for organic analyses, the meteorites of greatest interest to the organics community (e.g., carbonaceous chondrites, unequilibrated ordinary chondrites, and SNC meteorites) are, unfortunately, among the most valuable for other types of studies. Moreover, the curators and committees that are, in effect, the caretakers of these collections have to balance the interests of individual groups with the long-term preservation of material for future studies.

Thus, of the two models suggested for the coordination of studies of organic materials in meteorites, the more informal, community-based forum designed to set a consensus research agenda is likely to be more appropriate.

Recommendation: Plans should be developed for the establishment of an informal, community-based forum—modeled on the highly successful Mars Exploration Program Analysis Group (MEPAG)—charged to coordinate plans and develop priorities for the intensive investigation of the composition of organic materials in carbonaceous chondrites, SNC meteorites, and ordinary chondrites containing volatiles (including rare gases) that suggest relationships to the carbonaceous chondrites. The existing Curation and Analysis Planning Team for Extraterrestrial Materials (CAPTEM) might provide the seed from which such a community-based forum might be nurtured.

To provide comparability and to bring the best techniques to bear on each object, samples should be shared extensively between laboratories.

INCREASING THE SUPPLY OF METEORITES AVAILABLE FOR STUDY

The ready availability of and access to meteorites for laboratory studies, particularly the rare carbonaceous chondrites, is a key facet of the exploration of organic environments in the solar system. There are really only two ways to acquire meteorites:

- Collect them from areas of concentration, or
- Buy or trade for them from those who are lucky enough to see a fall or find an isolated specimen.

The pros and cons of these two strategies are considered below.

Collecting Meteorites

For researchers engaged in meteorite studies, the preferred means for acquiring samples is to collect them in the field. This approach has been far more productive in terms of numbers of meteorites and has led to major searches in those places in the world where meteorites are most likely to be found. Although meteorites fall to Earth in a random fashion, they are more likely to be spotted and collected if they fall in locations where they stand out against their background. In other words, a meteorite that falls in rocky terrain or dense vegetation is less conspicuous than one falling on a featureless ice sheet or salt flat. Thus, the hot and cold deserts of the world are the prime collecting areas.

Cold Deserts

Antarctica has been the world's most productive hunting ground for meteorites ever since 1969, when Japanese scientists discovered meteorite concentrations on bare ice stranding sites.[88] Since then some 35,000 meteorite specimens have been recovered by the combined activities of expeditions from the United States, Japan, China, the European EUROMET consortium, and other national programs. These meteorites are collected in probably the most sterile environment on Earth, and great pains are taken to minimize contamination in both their collection and curation.

The U.S. Antarctic Meteorite program is currently supported by the National Science Foundation's (NSF's) Office of Polar Programs, NASA, and the Smithsonian Institution. The field component of the program, the Antarctic Search for Meteorites (ANSMET), is currently supported by NSF and NASA. Initial examination and curation of samples are undertaken at the astromaterials curation facility at NASA's Johnson Space Center, and initial characterization and long-term curation are the responsibility of the Smithsonian's National Museum of Natural History.

The first ANSMET expedition (a joint U.S.-Japanese effort) discovered what turned out to be a significant concentration of meteorites at the Allan Hills in southern Victoria Land, a region that was to eventually reveal the famous martian meteorite, ALH 84001. In addition to ANSMET's long-standing NSF sponsorship, NASA has funded an expanded field effort, and this paid off with the U.S. team's recovery in December 2003 of the first martian meteorite in 9 years, as well as the collection of large numbers of carbonaceous chondrites.

Antarctica is such a productive source of meteorites not simply because rocks are easily spotted on icy surfaces, but also because the Antarctic environment actively concentrates meteorites in particular regions. That is, meteorites fall onto the ice and eventually become incorporated into the coastward flowing ice sheets. In those regions where ice flow is impeded by mountain ranges and subsurface obstructions, old deep ice can be forced to the surface. If this occurs in locations where strong winds ablate the surface ice, then meteorites that fell to Earth tens of thousands to millions of years ago will accumulate as a lag deposit. Meteorite densities of as much as one per square meter can be found in some locales.[89]

Hot Deserts

However, at the same time as the potential of Antarctica's cold deserts as a source of meteorites was being realized, other, smaller groups were exploring the potential of the world's hot deserts.[90] Although these environments lack the active concentration mechanism provided by ice flow and ablation in Antarctica, access to such regions is far less logistically challenging than mounting expeditions to Antarctica.

The Sahara has become a major source for meteorites. Many in the U.S. meteorite curation community have avoided dealing in these meteorites because of issues of unknown provenance and likely exportation from Libya. Some of these problems could be overcome with field searches, and the idea has been discussed (e.g., a joint Moroccan-U.S. field party). However, virtually all Saharan meteorites have been shown to have suffered considerable terrestrial weathering and have experienced much higher levels of contamination, and in many cases meteorite components have been replaced by carbon compounds (in this case, carbonates). In general there have been ample opportunities for the introduction or destruction of organic matter.

Hot Versus Cold Deserts

In considering collection of meteorites in hot as opposed to cold deserts, several factors have to be weighed. These include the relative cost per meteorite of collecting programs in different regions and the scientific utility of the samples collected. As productive as ANSMET is, the cost of supporting large collecting teams in remote locations is considerable in terms of both expense and utilization of scarce logistical resources (e.g., the limited supply of LC-135 transport aircraft that are required to support a broad range of Antarctic research activities, not just meteorite collecting). On the other hand, a typical desert collecting program, such as the highly successful Omani-Swiss program organized by Bern University and the Natural History Museum Bern,[91] requires logistical support in the form of four-wheel drive vehicles rather than specially equipped aircraft.

Some might argue that it is worthwhile to augment hot-desert collecting programs as an adjunct to the existing activities in Antarctica. Small-scale activities in new locales not only may result in an enhanced supply of scientifically interesting specimens, but also may well have collateral benefits such as jump-starting space-science-related activities in regions and nations not traditionally involved in such endeavors. Others will counter that contamination issues render meteorites collected in hot deserts less scientifically interesting, particularly for organic studies, than their antarctic counterparts.

Purchasing Meteorites

The attitudes of scientists toward the commercial meteorite trade are quite mixed. Many researchers believe that meteorites are a purely scientific legacy and that it is ethically no different to collect and trade in meteorites than it is to collect and trade in artifacts plundered from archeological sites. Some countries (but not the United States) have laws that protect meteorites from both commercial trade and export. Many scientists favor the adoption of similar legislation in the United States.

On the other hand, most researchers have neither the time nor the financial resources to find new meteorite collection areas and systematically search them for specimens (the exception being Antarctica). Thousands of meteorites available to researchers today would not have been found were it not for profit-driven collectors. The archeological-artifact analogy breaks down because, unlike, for example, an Anasazi pot, a meteorite can be divided into pieces without completely destroying its scientific value. Moreover, it can still be studied even completely out of the context of the location where it was originally found.

Thus, as a practical matter, most meteorite researchers recognize the reality of the commercial meteorite trade and have no choice but to rely on nonscientists who buy, sell, collect, and search for meteorites in order to obtain research material. Similarly, curators of collections in virtually every museum and university engage in commerce and trade with private meteorite dealers in order to obtain new material. Therefore, many meteorite researchers believe that a strong relationship between the commercial and scientific communities is an appropriate mechanism to maximize the return on the research-dollar.

The task group believes that serious consideration should be given to the selective acquisition of scientifically important meteorites. Much of the U.S. supply of the Murchison meteorite owes it existence to NASA funds made available to the Smithsonian Institution to buy a large, private collection shortly after the fall of Murchison. An interesting analogy exists today in Tagish Lake. Although small amounts of material have been made available to the scientific community through the Johnson Space Center, the bulk of the Tagish Lake fall remains in the hands of its finder, who has offered it for sale. The task group suggests that the greatest near-term scientific impact from a given expenditure of funds will result not from the enhancement of meteorite collecting programs but rather from the acquisition of a significant piece of the Tagish Lake meteorite so that it can be made available for study by the broader scientific community.

Recommendation: The scientific significance of the Tagish Lake meteorite is such that NASA, the National Science Foundation, the Smithsonian Institution, and other relevant organizations and agencies in the United States and their counterparts in Canada should examine the means by which a significant portion of this fall can be acquired, by purchase, exchange, or some other mechanism, so that samples can be made more widely available for study by the scientific community.

NOTES

1. J.J. Berzelius, "Found Humic Acid in Alais Carbonaceous Chondrite; Decided a Biological Origin Unlikely," *Annalen der Physik und Chemie* 33: 113, 1834.

2. M. Berthelot, "A Theoretical Paper Seeking to Explain Presence of Petroleumlike Hydrocarbons in Meteorites in Terms of a Reaction Between Metal Carbides and Water," *Comptes Rendus* 67: 849, 1868.

3. O. Botta and J.L. Bada, "Extraterrestrial Organic Compounds in Meteorites," *Surveys in Geophysics* 23: 411-467, 2002.

4. L. Becker, R.J. Poreda, and J.L. Bada, "Extraterrestrial Helium Trapped in Fullerenes in the Sudbury Impact Structure," *Science* 272: 249-252, 1996.

5. J.R. Cronin, S. Pizzarello, and D.P. Cruikshank, "Organic Matter in Carbonaceous Chondrites, Planetary Satellites, Asteroids and Comets," pp. 819-857 in *Meteorites and the Early Solar System* (J.F. Kerridge and M.S. Matthews, eds.), University of Arizona Press, Tucson, Ariz., 1988.

6. J.R. Cronin and S. Chang, "Organic Matter in Meteorites: Molecular and Isotopic Analyses of the Murchison Meteorite," pp. 209-258 in *The Chemistry of Life's Origin* (J.M. Greenberg, C.X. Mendoza-Gómez, and V. Pirronello, eds.), Kluwer Academic Press, The Netherlands, 1993.

7. P.G. Stoks and A.W. Schwartz, "Basic Nitrogen-heterocyclic Compounds in the Murchison Meteorite," *Geochimica et Cosmochimica Acta* 46: 309-315, 1982.

8. J.R. Cronin, S. Pizzarello, and D.P. Cruikshank, "Organic Matter in Carbonaceous Chondrites, Planetary Satellites, Asteroids and Comets," pp. 819-857 in *Meteorites and the Early Solar System* (J.F. Kerridge and M.S. Matthews, eds.), University of Arizona Press, Tucson, Ariz., 1988.

9. Some nonbiological amino acids that have a small excess of one stereoisomer have been detected in the Murchison meteorite. These are believed to have been formed abiotically. See the section "Chirality: Enantiomeric Excesses Observed in Stereoisomers from Carbonaceous Chondrites" in this chapter.

10. O. Botta and J.L. Bada, "Extraterrestrial Organic Compounds in Meteorites," *Surveys in Geophysics* 23: 411-467, 2002.

11. M.A. Septhon, "Organic Compounds in Carbonaceous Meteorites," *Natural Product Reports* 19: 292-311, 2002.

12. J.R. Cronin and S. Chang, "Organic Matter in Meteorites: Molecular and Isotopic Analyses of the Murchison Meteorite," pp. 209-258 in *The Chemistry of Life's Origin* (J.M. Greenberg, C.X. Mendoza-Gómez, and V. Pirronello, eds.), Kluwer Academic Press, The Netherlands, 1993.

13. G. Cooper, W. Onwo, and J.R. Cronin, "Alkyl Phosphonic Acids and Sulfonic Acids in the Murchison Meteorite," *Geochimica et Cosmochimica Acta* 56: 4109-4115, 1992.

14. L. Becker, R.J. Poreda, and T.E. Bunch, "Fullerenes: An Extraterrestrial Carbon Carrier Phase for Noble Gases," *Proceedings of the National Academy of Sciences* 97(7): 2979-2983, 2000.

15. O. Botta and J.L. Bada, "Extraterrestrial Organic Compounds in Meteorites," *Surveys in Geophysics* 23: 411-467, 2002.

16. G.W. Cooper, N. Kimmich, W. Belisle, J. Sarinana, K. Brabham, and L. Garrel, "Carbonaceous Meteorites as a Source of Sugar-related Organic Compounds for the Early Earth," *Nature* 414: 879-883, 2001.

17. J.R. Cronin and S. Chang, "Organic Matter in Meteorites: Molecular and Isotopic Analyses of the Murchison Meteorite," pp. 209-258 in *The Chemistry of Life's Origin* (J.M. Greenberg, C.X. Mendoza-Gómez, and V. Pirronello, eds.), Kluwer Academic Press, The Netherlands, 1993.

18. J.R. Cronin, S. Pizzarello, and D.P. Cruikshank, "Organic Matter in Carbonaceous Chondrites, Planetary Satellites, Asteroids and Comets," pp. 819-857 in *Meteorites and the Early Solar System* (J.F. Kerridge and M.S. Matthews, eds.), University of Arizona Press, Tucson, Ariz., 1988.

19. J.R. Cronin and S. Chang, "Organic Matter in Meteorites: Molecular and Isotopic Analyses of the Murchison Meteorite," pp. 209-258 in *The Chemistry of Life's Origin* (J.M. Greenberg, C.X. Mendoza-Gómez, and V. Pirronello, eds.), Kluwer Academic Press, The Netherlands, 1993.

20. O. Botta and J.L. Bada, "Extraterrestrial Organic Compounds in Meteorites," *Surveys in Geophysics* 23: 411-467, 2002.

21. J.R. Cronin and S. Chang, "Organic Matter in Meteorites: Molecular and Isotopic Analyses of the Murchison Meteorite," pp. 209-258 in *The Chemistry of Life's Origin* (J.M. Greenberg, C.X. Mendoza-Gómez, and V. Pirronello, eds.), Kluwer Academic Press, The Netherlands, 1993.

22. R.F. Curl, "On the Formation of Fullerenes," *Philosophical Transactions of the Royal Society of London: Physical Sciences and Engineering* 343: 19-32, 1993.

23. S. Pizzarello, Y. Huang, L. Becker, R.J. Poreda, R.A. Nieman, C. Cooper, and M. Williams, "Organic Matter in Tagish Lake Carbonaceous Chondrite," *Science* 293: 2236-2239, 2001.

24. P. Ehrenfreund, D. Glavin, O. Botta, G.W. Cooper, and J.B. Bada, "Extraterrestrial Amino Acids in Orgueil and Ivuna: Tracing the Parent Body of CI-type Carbonaceous Chondrites," *Proceedings of the National Academy of Sciences* 98: 2138-2141, 2001.

25. G.E. Folsome, J.G. Lawless, M. Ramirez, and C. Pomnamperuma, "Heterocyclic Compounds Indigenous to the Murchison Meteorite," *Nature* 232: 108-109, 1971.

26. G.E. Folsome, J.G. Lawless, M. Ramirez, and C. Pomnamperuma, "Heterocyclic Compounds Recovered from Carbonaceous Chondrites," *Geochimica et Cosmochimica Acta* 46: 455-465, 1973.

27. R. Hayatsu, M.H. Studier, L.P. Moore, and E. Anders, "Purines and Triazines in the Murchison Meteorite," *Geochimica et Cosmochimica Acta* 39: 471-478, 1975.

28. W. Van der Velden and A.W. Schwartz, "Search for Purines and Pyrimidines in the Murchison Meteorite," *Geochimica et Cosmochimica Acta* 41: 961-968, 1977.

29. P.G. Stoks and A.W. Schwartz, "Nitrogen-heterocyclic Compounds in Meteorites: Significance and Mechanism of Formation," *Geochimica et Cosmochimica Acta* 45: 563-569, 1981.

30. P.G. Stoks and A.W. Schwartz, "Basic Nitrogen-heterocyclic Compounds in the Murchison Meteorite," *Geochimica et Cosmochimica Acta* 46: 309-315, 1982.

31. S. Pizzarello, Y. Huang, L. Becker, R.J. Poreda., R.A. Nieman, C. Cooper, and M. Williams, "Organic Matter in Tagish Lake Carbonaceous Chondrite," *Science* 293: 2236-2239, 2001.

32. E.T. Peltzer and J.L. Bada, "alpha-Hydroxycarboxylic Acids in the Murchison Meteorite," *Nature* 272: 443-444, 1978.

33. P.G. Stoks and A.W. Schwartz, "Basic Nitrogen-heterocyclic Compounds in the Murchison Meteorite," *Geochimica et Cosmochimica Acta* 46: 309-315, 1982.

34. R. Hayatsu, M.H. Studier, L.P. Moore, and E. Anders, "Purines and Triazines in the Murchison Meteorite," *Geochimica et Cosmochimica Acta* 39: 471-478, 1975.

35. J.P. Ferris and W.J. Hagan, Jr., "HCN and Chemical Evolution," *Tetrahedron* 40: 1093-1120, 1984.

36. S.L. Miller and H.C. Urey, "Organic Compound Synthesis on the Primitive Earth," *Science* 130: 245-251, 1959.

37. C.C. Kung, R. Hayatsu, M.H. Studier, and R.N. Clayton, "Nitrogen Isotope Fractionations in the Fischer-Tropsch Synthesis and the Miller-Urey Reaction," *Earth and Planetary Science Letters* 46: 141-146, 1979.

38. P.G. Stoks and A.W. Schwartz, "Nitrogen-heterocyclic Compounds in Meteorites: Significance and Mechanism of Formation," *Geochimica et Cosmochimica Acta* 45: 563-569, 1981.

39. S. Pizzarello, Y. Huang, L. Becker, R.J. Poreda., R.A. Nieman, C. Cooper, and M. Williams, "Organic Matter in Tagish Lake Carbonaceous Chondrite," *Science* 293: 2236-2239, 2001.

40. S. Pizzarello, Y. Huang, L. Becker, R.J. Poreda., R.A. Nieman, C. Cooper, and M. Williams, "Organic Matter in Tagish Lake Carbonaceous Chondrite," *Science* 293: 2236-2239, 2001.

41. J.M. Hayes, "Organic Constituents of Meteorites—A Review," *Geochimica et Cosmochimica Acta* 31: 1395-1440, 1967.

42. E. Zinner, "Interstellar Cloud Material in Meteorites," pp. 956-983 in *Meteorites and the Early Solar System* (J.F. Kerridge and M.S. Mathews, eds.), University of Arizona Press, Tucson, Ariz., 1988.

43. R. Hayatsu, R.E. Winans, R.G. Scott, L.P. Moore, and M.H. Studier, "Phenolic Ethers in the Organic Polymer of the Murchison Meteorite," *Science* 207: 1202-1204, 1980.

44. M.A. Sephton, C.T. Pillinger, and I. Gilmour, "Small-scale Hydrous Pyrolysis of Macromolecular Material in Meteorites," *Planetary and Space Science* 47: 181-187, 1999.

45. J.R. Cronin, S. Pizzarello, and J.S. Frye, "^{13}C NMR Spectroscopy of the Insoluble Carbon of Carbonaceous Chondrites," *Geochimica et Cosmochimica Acta* 51: 299-303, 1987.

46. A. Gardinier, S. Derenne, F. Robert, F. Behar, C. Largeau, and J. Maquet, "Solid State CP/MAS ^{13}C NMR of the Insoluble Organic Matter of the Orgueil and Murchison Meteorites: Quantitative Study," *Earth Planetary Science Letters* 184: 9-21, 2000.

47. G.D. Cody, C.M.O'D. Alexander, and F. Tera, "Solid State (^{1}H and ^{13}C) Nuclear Magnetic Resonance Spectroscopy of Insoluble Organic Residue in the Murchison Meteorite: A Self-consistent Quantitative Analysis," *Geochimica et Cosmochimica Acta* 66: 1851-1865, 2002.

48. S. Pizzarello, Y. Huang, L. Becker, R.J. Poreda., R.A. Nieman, C. Cooper, and M. Williams, "Organic Matter in Tagish Lake Carbonaceous Chondrite," *Science* 293: 2236-2239, 2001.

49. G.D. Cody, C.M.O'D. Alexander, and F. Tera, "Compositional Trends in Chondritic Organic Solids Within and Between Meteoritic Groups," Lunar and Planetary Science Conference XXXIV, Abstract 1822, 2003.

50. G.D. Cody, C.M.O'D. Alexander, and F. Tera, "Compositional Trends in Chondritic Organic Solids Within and Between Meteoritic Groups," Lunar and Planetary Science Conference XXXIV, Abstract 1822, 2003.

51. G.D. Cody, C.M.O'D. Alexander, and F. Tera, "Solid State (^{1}H and ^{13}C) Nuclear Magnetic Resonance Spectroscopy of Insoluble Organic Residue in the Murchison Meteorite: A Self-consistent Quantitative Analysis," *Geochimica et Cosmochimica Acta* 66: 1851-1865, 2002.

52. E. Anders and E. Zinner, "Interstellar Grains in Primitive Meteorites: Diamond, Silicon Carbide, and Graphite," *Meteoritics* 28: 490-514, 1993.

53. O. Botta and J.L. Bada, "Extraterrestrial Organic Compounds in Meteorites," *Surveys in Geophysics* 23: 411-467, 2002.

54. P. Ehrenfreund, W. Irvine, L. Becker, J. Blank, J.R. Brucato, L. Colangeli, S. Derenne, D. Despois, A. Dutrey, H. Fraaije, A. Lazcano, T. Owen, and F. Robert, "Astrophysical and Astrochemical Insights into the Origins of Life," *Reports on Progress in Physics* 65: 1427-1487, 2002.

55. C.M.O'D. Alexander, S.S. Russell, J.W. Arden, R.D. Ash, M.M. Grady, and C.T. Pillnger, "The Origin of Chondritic Macromolecular Organic Matter: A Carbon and Nitrogen Isotope Study," *Meteoritics and Planetary Science* 33: 603-622, 1998.

56. M.A. Sephton, C.T. Pillinger, and I. Gilmour, "Aromatic Moieties in Meteoritic Macromolecular Materials: Analyses by Hydrous Pyrolysis and δ ^{13}C of Individual Compounds," *Geochimica et Cosmochimica Acta* 64: 321-328, 2000.

57. M.A. Sephton, C.T. Pillinger, and I. Gilmour, "Aromatic Moieties in Meteoritic Macromolecular Materials: Analsyses by Hydrous Pyrolysis and δ ^{13}C of Individual Compounds," *Geochimica et Cosmochimica Acta* 64: 321-328, 2000.

58. C.M.O'D. Alexander, S.S. Russell, J.W. Arden, R.D. Ash, M.M. Grady, and C.T. Pillnger, "The Origin of Chondritic Macromolecular Organic Matter: A Carbon and Nitrogen Isotope Study," *Meteoritics and Planetary Science* 33: 603-622, 1998.

59. S.L. Miller and H.C. Urey, "Organic Compound Synthesis on the Primitive Earth," *Science* 130: 245-251, 1959.

60. N.P. Chang, J. Leon, and J.P. Mercader, "Kahler Manifolds with Vanishing Chiral Potential," *Physical Review Letters* 58: 1505-1507, 1987.

61. W.A. Bonner, "Homochirality and Life," pp. 159-188 in *D-Amino Acids in Sequences of Secreted Peptides of Multicellular Organisms* (P. Jolles, ed.), Birkhauser Verlag, Basel, Switzerland, 1998.

62. M.H. Engel and S.A. Macko, "Isotopic Evidence for Extraterrestrial Non-racemic Amino Acids in the Murchison Meteorite," *Nature* 389: 265-268, 1997.

63. J.R. Cronin and S. Pizzarello, "Enantiomeric Excesses in Meteoritic Amino Acids," *Science* 275: 951-955, 1997.

64. S. Pizzarello and J.R. Cronin, "Alanine Enantiomers in the Murchison Meteorite," *Nature* 394: 236, 1998.

65. S. Pizzarello and J.R. Cronin, "Non-racemic Amino Acids in the Murray and Murchison Meteorites," *Geochimica et Cosmochimica Acta* 64: 329-338, 2000.

66. O. Vandenabeele-Trambouze, M. Dobrijevic, D. Despois, A. Commeyras, M. Geffard, C. Bayle, M. Albert, and M.F. Grenier-Loustalot, "Amino-acids Enantiomeric Ratio Determination in Micrometeorites: Analytical Development and Interest for Mars Samples," *Frontiers of Life—XIIth Rencontres de Blois* (L.M. Celnikier and J. Tran Thanh Van, eds.), Proceedings of the XIIth Rencontres de Blois: Frontiers of Life Conference, June 25-1 July 1, 2000, Chateau de Blois, France, The Gioi Publishers, Hanoi, Vietnam, 2003.

67. W. Thiemann and U. Meierhenrich, "ESA Mission ROSETTA Will Probe for Chirality of Cometary Amino Acids," *Origins of Life and Evolution of the Biosphere* 31: 199-210, 2001.

68. J.R. Cronin and S. Pizzarello, "Enantiomeric Excesses in Meteoritic Amino Acids," *Science* 275: 951-955, 1997.

69. J. Bailey, A. Chrysostomou, J.H. Hough, T.M. Gledhill, A. McCall, S. Clark, F. Ménard, and M. Tamura, "Circular Polarization in Star-formation Regions: Implications for Biomolecular Homochirality," *Science* 281: 672-674, 1998.

70. E. Rubenstein, W.A. Bonner, H.P. Noyes, and G.S. Brown, "Supernovae and Life," *Nature* 300: 118, 1983.

71. P. Ehrenfreund, M.P. Bernstein, J.P. Dworkin, S.A. Sandford, and L.J. Allamandola, "The Photostability of Amino Acids in Space," *Astrophysical Journal Letters* 550: 95-99, 2001.

72. C.M.O'D. Alexander, S.S. Russell, J.W. Arden, R.D. Ash, M.M. Grady, and C.T. Pillnger, "The Origin of Chondritic Macromolecular Organic Matter: A Carbon and Nitrogen Isotope Study," *Meteoritics and Planetary Science* 33: 603-622, 1998.

73. I.R. Kaplan, E.T. Degens, and J.H. Reuter, "Organic Compounds in Stony Meteorites," *Geochimica et Cosmochimica Acta* 27: 805-834, 1963.

74. For a review see, for example, J.M. Hayes, "Organic Constituents of Meteorites—A Review," *Geochimica et Cosmchimica Acta* 31: 1395-1440, 1967.

75. M.A. Sephton, C.T. Pillinge, and I. Gilmour, "Normal Alkanes in Meteorites: Molecular ^{13}C Values Indicate an Origin by Terrestrial Contamination," *Precambrian Research* 106: 45-56, 2001.

76. S.J. Clemett, C.R. Maechling, R.N. Zare, and C.M.O'D Alexander, "Analysis of Polycyclic Aromatic Hydrocarbons in Seventeen Ordinary and Carbonaceous Chondrites," *Lunar and Planetary Science* 23: 233-234, 1992.

77. L.J. Kovalenko, C.R. Maechling, S.J. Clemett, J.-M. Philipposz, R.N. Zare, and C.M.O'D. Alexander, "Microscopic Organic Analysis Using Two-step Laser Mass Spectrometry: Application to Meteoritic Acid Residues," *Analytical Chemistry* 64: 682-690, 1992.

78. M.A. Sephton, C.T. Pillinger, and I. Gilmour, "Evidence from the Isotopic Compositions of Individual Molecules for the Indigeneity of PAH in Meteorites," Lunar and Planetary Science Conference XXVIII, Abstract 1732, 1999.

79. See, for example, H.Y. McSween, Jr., and A.H. Treiman, "Martian Meteorites," Chapter 6 of *Planetary Materials: Reviews in Mineralogy*, Volume 36 (J.J. Papike ed.), Mineralogical Society of America, Chantilly, Va., 1998.

80. See, for example, the Meteorite Catalogue Database maintained by the U.K.'s Natural History Museum, available online at http://internt.nhm.ac.uk/jdsml/research-curation/projects/metcat/, last accessed January 19, 2007; or the Mars Meteorite Compendium maintained at NASA's Johnson Space Center, available online at http://www-curator.jsc.nasa.gov/curator/antmet/mmc/index.cfm, last accessed January 19, 2007.

81. I.P. Wright, R.H. Carr, and C.T. Pillinger, "Carbon Abundances and Isotopic Studies of Shergotty and Other Shergottite Meteorites," *Geochimica et Cosmochimica Acta* 50: 983-991, 1986.

82. I.P. Wright, M.M. Grady, and C.T. Pillinger, "Organic Materials in a Martian Meteorite," *Nature* 340: 220-222, 1989.

83. M.A. Sephton, I.P. Wright, I. Gilmour, J.W. de Leeuw, M.M. Grady, and C.T. Pillinger, "High Molecular Weight Organic Matter in Martian Meteorites," *Planetary and Space Science* 50: 711-716, 2002.

84. D.S. McKay, E.K. Gibson, Jr., K.L. Thomas-Keprta, H. Vali, C.S. Romanek, S.J. Clemett, X.D.F. Chillier, C.R. Maechling, and R.N. Zare, "Search for Past Life on Mars: Possible Relic Biogenic Activity in Martian Meteorite ALH 84001," *Science* 273: 924-930, 1996.

85. L. Becker, B. Popp, T. Rust, and J.L. Bada, "The Origin of Organic Matter in the Martian Meteorite ALH 84001," *Earth and Planetary Science Letters* 167: 71-79, 1999.

86. A.J.T. Jull, C. Courtney, D.A. Jeffrey, and J.W. Beck, "Isotopic Evidence for a Terrestrial Source of Organic Compounds Found in Martian Meteorites: Allan Hill 84001 and Elephant Moraine 79001," *Science* 279: 366, 1999.

87. For more information about MEPAG see http://mepag.jpl.nasa.gov/. Last accessed January 19, 2007.

88. M. Yoshida, H. Ando, K. Omoto, R. Naruse, and Y. Ageta, "Discovery of Meteorites Near Yamato Mountains, East Antarctica," *Antarctic Record* 39: 62-65, 1971.

89. For a summary of Antarctic meteorite field work, overviews of meteoritic and related glaciological investigations, and discussions of concentration mechanisms see, for example, W.A. Cassidy, R.P. Harvey, J. Schutt, G. Delisle, and K. Yanai, "The Meteorite Collection Sites of Antarctica," *Meteoritics* 27: 490-525, 1992.

90. See, for example, A. Bischoff, "Fantastic New Chondrites, Achondrites, and Lunar Meteorites as the Result of Recent Meteorite Search Expeditions in Hot and Cold Deserts," *Earth, Moon, and Planets* 85-86: 87-95, 2001.

91. B.A. Hofmann, E. Gnos, A. Al-Kathiri, H. Al-Azri, and A. Al-Murazza, "Omani-Swiss Meteorite Search 2001-2003 Project Overview," *Meteoritics and Planetary Science* 38: Abstract 5093.

4

Primitive Bodies

Primitive bodies are those bodies that have undergone the least amount of change (chemically, thermally, and physically) since their formation in the solar nebula. These objects include main belt asteroids, Trojan asteroids, small satellites, comets, Centaurs,[1] and Kuiper belt objects (KBOs).[2] Because of their small size, and their often low albedos and large distances from the Sun and Earth that contribute to their faintness, the observation and understanding of their organic components are challenging.

FORMATION AND DYNAMICAL REGIMES

Asteroids orbit the Sun mainly in the region between Mars and Jupiter and represent inner solar system planetesimals that were unable to accrete into a planet because of the gravitational perturbations of Jupiter.

Comets are icy bodies that probably formed primarily in the Uranus-Neptune region and beyond (from ~20 to 50 AU); however, some of the material may have formed as close to the Sun as Jupiter. Many of the comets formed near Uranus-Neptune (~20 AU) were perturbed outward in what is now known as the Oort cloud, via gravitational interactions with the giant planets. Subsequent stellar perturbations can inject these Oort cloud comets into the inner solar system where they may become visible as either short- or long-period comets. Objects that originally formed farther out, in the vicinity of the Kuiper belt (30 to 50 AU), are sometimes perturbed inward and become short-period comets. As they migrate inward, they are observed as Centaurs. The importance of the different formation locations is that bodies from different dynamical regimes will have formed at different temperatures, will have had different thermal histories, and may, therefore, have different chemistries. The small outer planet satellites and Trojan asteroid populations probably represent planetesimals that formed in the outer solar system near the orbit of Jupiter and were subsequently captured into dynamically stable niches,[3] and thus may share similarities with other outer solar system icy bodies such as KBOs, Centaurs, and comets. The interiors of Trojan asteroids are expected to be rich in H_2O and other volatile materials.

The precursor cometary material may be a combination of unaltered interstellar grains/ices and volatiles that would have sublimated or vaporized as the material fell through accretion shock fronts in the protoplanetary disk, and then subsequently recondensed. Water ice can exist in either an amorphous or a crystalline phase depending on its condensation temperature and thermal history. If water ice condenses from the gas phase at temperatures below 100 K, then it will not have enough energy for an orderly crystalline structure and will be present in the amorphous phase. Laboratory experiments have shown that amorphous ice has the ability to trap large amounts of gases in spaces between the water molecules,[4] with gas/ice ratios as high as 3 to 3.5 (by number) at 30 K. As the temperature of condensation increases above 30 K the amount of gas that can be trapped drops quickly by 6 orders of magnitude at the temperature of 100 K, and different species are preferentially trapped, resulting in fractionation. When the ices are warmed, the ice structure reorganizes and the spaces between the molecules get smaller (annealing), and volatiles can be released. In addition, the amorphous ice will undergo an exothermic phase transition to its crystalline form, peaking near 137 K, also resulting in the release of trapped volatiles. Once the phase transition has occurred, the ice will not revert to the amorphous form if the temperature is lowered, unless there is some process (such as cosmic-ray irradiation, or vaporization and recondensation at low temperature) that destroys the structure in the ice.

MEASUREMENT TECHNIQUES—ORGANICS

The compositions of the primitive, low-albedo bodies are only broadly understood, whereas researchers have more information for the gaseous comas (the cloud of gas and dust that extends from the nucleus) of comet nuclei. Comets have been measured photometrically in the optical and spectroscopically in the near-infrared, and volatiles escaping from comet nuclei have been observed at radio wavelengths. In addition, several space missions have made in situ measurements. The materials expected on the surfaces of these bodies include minerals, and with increasing distance from the Sun water-ice, other volatiles, and solid organic materials.

While optical spectroscopy has been used to identify dissociation fragments (daughters) in cometary comas for nearly 100 years, it is often difficult to deduce their parent species. The inherent difficulty is that the volatiles sublime from the nucleus and shortly after entering the coma can experience photolysis and photodissociation. Many species can be seen as both a nuclear (parent) and a dissociation product. Models are then used to infer from the observed spectroscopic band strengths the abundance of various parent or original species coming from the nucleus. One of the surprising things learned from spectroscopic observations of recent bright comets, such as C/1995 O1 (Hale Bopp), is that it is very likely that chemical reactions may have been occurring in the inner coma, and thus these types of observations may not reflect the primordial composition in the cometary interior and may not be entirely useful indicators of the preservation of interstellar material. With recently observed bright comets such as C/1995 O1 (Hale Bopp), C/1996 B2 (Hyakutake), and C/1999 H1 (Lee), it was possible to make direct observations of the parent molecules in the infrared and radio wavelengths.

Spectroscopy of the solid bodies is more difficult owing to their faintness, making it very expensive (i.e., requiring considerable time) in terms of telescope time. In the optical wavelength region the most prominent spectral features come from mineral assemblages, whereas the signatures of organic material lie in the near-infrared region between 1 and 5 μm. Water ice also has strong solid-state absorption bands in this region, as do other volatiles of interest. Good near-infrared spectrometers have been installed only recently on large telescopes to facilitate the search for organic signatures in this wavelength region. The identification of organic species in the spectra requires the use of spectral models that apply Hapke scattering theory using the complex refractive indices of minerals, volatiles, and organic materials in combination to match the features in the spectrum. One of the problems is that the models are very complex and non-unique. In addition, optical constants (complex refractive indices), which must be measured in the laboratory, are known for only a relatively small number of the compounds that are likely to be present.

SYNTHESIS AND DESTRUCTION OF ORGANIC MATERIALS

Organic molecules can be both synthesized and destroyed on outer solar system solids by irradiation. The surface materials on small bodies or grains are typically exposed to charged-particle and ultraviolet radiation, producing radiolysis and photolysis. There is abundant laboratory evidence that carbon-containing frozen mixtures will form complex organics when exposed to radiation and that complex organics break down under irradiation. Under the influence of radiation, these surfaces will grow progressively redder and darker as hydrogen is lost.[5-7] However, there is as yet no definitive observational evidence that radiation processing plays an important role in forming organics on solar system surfaces (note that there is abundant evidence now for radiation processing).

Although there is considerable speculation that the red-spectral character of many outer solar system surfaces is due to organics and/or to the processing of these materials, again there is no direct evidence. Determining if such processes occur is important, because exposure to radiation affects the reflectance properties of optical surfaces, which in turn impacts observations made by remote sensing.

The laboratory evidence for chemistry induced by radiolysis and photolysis in low-temperature solids is clear. However, detecting chemical products in a solid by remote sensing is more difficult than it is for the gas phase. As a result, there has been much less progress in both understanding solid-state radiation chemistry and applying that to remote sensing observations. While there has been considerable improvement in reflectance studies of outer solar system surfaces, it is now realized that vertical mixing does occur on geologically "young" surfaces like that of Europa. It is also clear that tenuous atmospheres on small objects, as well as certain gas-phase species in the interstellar medium, are produced by stimulated desorption and sputtering. Therefore, it is now important to understand the chemistry induced by the radiolysis and photolysis of ice containing a variety of carbon species in the solid state. Further, with the developing plans to explore organic environments in the outer solar system, there is a need to understand the effect of radiation on any indigenous or delivered organics.

As in the gas phase, the chemical reactions of interest in solids are driven primarily by bond breaking and ionization due to ultraviolet photons and energetic charged particles. This process creates free radicals (as well as ions and electrons) that can react with neighboring molecules. The resulting chemistry is different from that in the gas-phase chemistry in three significant ways. First, the reactive species produced are often immobile, leading to enhanced geminate recombination and hot-atom/insertion interactions. As trapped species (including electrons) are often stable for long periods, a later heat pulse or further irradiation may be needed to drive reactions. A second difference from gas-phase chemistry is that energetic charged particles produce a locally dense set of excitations in a solid. Therefore, the chemical pathways can be very different from those in photolysis of a gas or a solid and different from charged-particle irradiation in a gas. Finally, because solid objects with low gravity lose their hydrogen preferentially, the surfaces of interest are typically oxidizing, so that the organic chemistry occurring there can differ significantly from that in hydrogen-dominated regions of space.

In the outer solar system galactic cosmic rays, solar particles (solar wind and solar energetic particles), charged particles trapped in the magnetic fields of the giant planets, and solar ultraviolet and extreme-ultraviolet radiation are all of interest because each is capable of providing energy for radiolysis or photolysis to take place on these surfaces. These various forms of radiation act over very different time scales and depths into a surface. Typically the relevant time scales increase with depth into a solid. That is, on a stable surface the optical layer is often modified in periods of months to hundreds of years, whereas depths of the order of centimeters at unit density are modified in millions of years and depths of the order of a meter are modified in the lifetime of the solar system. In addition, there is evidence in the early solar system for intense solar particle radiation near the edge of the accretion disk or during the Sun's T-Tauri phase, so that materials incorporated into objects, particularly refractory organics, may have been pre-irradiated.

The major sources of carbon in interstellar ices are CO, CO_2, and CH_3OH and, possibly, condensed carbon, PAHs, or fullerenes. In contrast, in the vicinity of giant-planet formation CH_4 is a principal source of carbon, as is the case in the solar nebula where chemical equilibrium has occurred. Therefore, the thermal and radiation processing of these species in the presence of H_2O, NH_3, N_2, and others, can, in principle, lead to organic chemical pathways on planetary satellites that are different from those on relatively unprocessed bodies, such as Kuiper belt objects. However, even if different initial chemical conditions occurred, the exchange of materials by impacts and by plasma transport can homogenize the cometary bodies and icy satellites. Therefore, a key to understanding the evolution of outer solar system objects is to understand their organic inventory in detail, particularly the trace organic molecules, and to understand the inventory as a function of distance from the Sun. A critical question is whether there is any evidence, other than the deuterium/hydrogen ratios, that any outer solar system body contains unprocessed interstellar organics.

Comets and Centaurs can have highly elliptical orbits and thus can pass through very different thermal environments. This environmental variability leads to significant evolution in their interiors and possible chemical processing. While residing in the distant reaches of the solar system, comets will undergo irradiation of the surface layers, and in addition may be heated to at least 30 K to a depth of 20 to 60 m by the passage of luminous stars, and it is believed that most comets may have been heated as high as 45 K to a depth of 1 m from stochastic supernova events. This heating may result in depletion of volatiles in the upper layers. The evolution of meteorites provides evidence of an early heat source in the solar system; while many researchers believe that radioactive ^{26}Al was likely responsible for radiogenic heating of large bodies, strong arguments can be made that other radionuclides (e.g., ^{60}Fe) also played important roles. Prialnik et al. have examined the role of ^{26}Al in the possible evolution of cometary interiors.[8] Because there is evidence for amorphous ice in comets, they cannot have been heated above 137 K. However, for larger comet nuclei, such as Kuiper belt objects, models have shown that it is possible under certain interior conditions (which include specific porosities, the fraction of other volatiles present, and so on) to have melting in the interior creating reservoirs of liquid water.[9] This possibility has interesting implications in view of recent experiments on organic compounds that show self-organizing behavior in the presence of liquid water. During the active phase, when a comet passes within the inner solar system and experiences significant solar heating, the comet's surface and subsurface layers (up to a few meters in depth) will be depleted in volatile material and also may lose the highly volatile radicals that were created from galactic cosmic-ray processing.

The phase transition and subsequent gas release are sensitive to the physical and chemical properties of the nucleus, as well as the orbital evolution of the comet. As volatiles are depleted when heated, a dust mantle will form on the surface, which may erode during periods of high activity and the thickness of which will be a function of orbital evolution (Figure 4.1). These types of effects have been observed in comets, ranging from production of vastly different amounts of volatile material between comets coming close to the Sun for the first time and comets that have

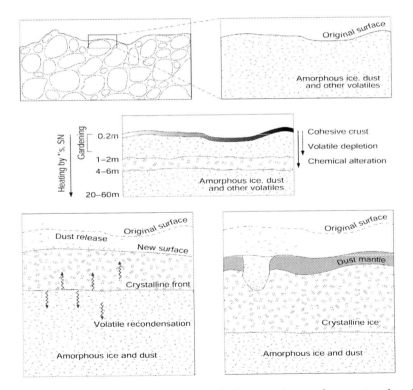

FIGURE 4.1 Diagram showing the sequence of aging processes in the upper layers of a comet nucleus from (a) the pristine state, consisting of primordial planetesimals (upper left, enlargement at upper right), (b) to the alterations it undergoes while stored in the Oort cloud including a possible crystalline core caused by radioactive heating from ^{26}Al (central diagram) to (c) the changes in the surface during the active phase (lower left) and (d) near the end of its evolution as a dust mantle builds up (lower right). SOURCE: K.J. Meech, "Physical Aging of Comets," pp. 195-210 in *Evolution and Source Regions of Asteroids and Comets* (J. Svoren, E.M. Pittich, and H. Rickman, eds.), IAU colloquium No. 173, Tatranska Lomnica, Slovakia, August 24-28, 1998, Astronomical Institute of the Slovak Academy of Sciences, Tatranska Lomnica, The Slovak Republic, 1999.

spent a large amount of time in the inner solar system, to observations of carbon-chain species (e.g., C_2, C_3) in comets dynamically associated with Jupiter. It has been suggested that some process in the solar nebula may have preferentially produced or destroyed the carbon chain molecules at the distance of the Kuiper belt, the source region for most of the Jupiter-family comets,[10] although observationally researchers cannot yet distinguish this from an evolutionary effect in comets. Therefore, comets are unusual because of their passage through very different thermal regimes, their possible preservation of pre-solar materials, and their small sizes, all of which create different physical and chemical processing conditions compared to those associated with planets and their major satellites.

ORGANIC INVENTORY

Surfaces of small primitive bodies are composed of varying fractions of refractory material and frozen volatiles. Many of these surfaces, in particular in the outer solar system, exhibit abundant evidence for dark materials that are generally spectrally neutral (i.e., they are equally bright at visual, red, and infrared wavelengths) to red (i.e., they are brighter at red and, in particular, infrared wavelengths than they are at visual wavelengths). The redness of these objects has led to the suggestion that their surfaces contain organics or a carbon-rich material, possibly frozen in an icy matrix, and therefore may be of interest in studies of abiotic chemical processes. However, the presence of specific organics on most of these bodies remains speculative because of the lack of large numbers of spectra and because models are limited by lack of laboratory measurements.

Asteroids

Asteroids can be classified by their surface spectral properties, colors, and albedos. S-type asteroids have relatively high albedos, significant spectral features associated with anhydrous silicates, and a general "reddish" spectral slope.[11] These are primarily assemblages of anhydrous silicates, NiFe metal and iron sulfide, analogous to ordinary chondrites, primitive achondrites, stony-iron meteorites, and silicate-bearing iron meteorites. C-type asteroids tend to have a very low albedo and are neutral in color, with subdued features indicating the presence of phyllosilicates (clay minerals). Most C-type asteroids appear analogous to CI1- or CM2-type carbonaceous chondrites. Other types of asteroids have been defined, including the common P- and D-types in the outer parts of the asteroid belt, but little is known about their composition and origin. D-type objects have low albedos and are characterized by featureless spectra with neutral to slightly red slopes shortward of 0.55 μm, and very red slopes longward of 0.55 μm.[12] S-class asteroids are those that in general are closest to the Sun and contain little carbon, while the C-type asteroids (based on their meteorite analogs) have carbon compounds that have been transformed by heat and water. The more distant P- and D-type asteroids do not appear to have been heated, and so their organics have not been subjected to alteration by high temperatures and liquid water. Presumably the materials in the P- and D-type asteroids contain organics whose structures are similar to those in comets. The C-, P-, and D-type asteroids are believed to be the source of both the carbonaceous meteorites and a large fraction of the carbon-containing interplanetary dust particles deposited on the inner planets. Indeed, the reflectance spectrum of the Tagish Lake meteorite is similar to the observed spectra of D-type asteroids.[13] On the other hand, there has been no direct observation of organic material on any class of asteroids in the main belt. Among the outer belt asteroids and the Trojans there is no evidence for ices, and even evidence for specific minerals and organic material is indirect. However, some researchers have argued that the only plausible material that can provide both the low albedo and the spectrally red slope of their reflectivities is complex organic solid material.[14] Caution is needed, however, because albedo and color are highly nonlinear functions of composition, grain size, and shapes (see the section "Laboratory Opportunities" below in this chapter).

Comets Within the Inner Solar System

The first analog for the composition of comets began with the "dirty snowball" model of Whipple.[15] Whipple proposed that cometary nuclei were made predominantly of water ice with an admixture of rocky particles and large organic entities or CHON (carbon, hydrogen, oxygen, and nitrogen) particles.[16,17] The chemical composition of comet nuclei has been constrained by Greenberg, who proposed that 26 percent of the mass is incorporated in silicates, 23 percent in refractory material, 9 percent in small carbonaceous molecules, and the remainder in water ice and trace gases (e.g., CO, CH_4, C_2H_2, and others).[18] These trace elements are important, however, as they provide clues about the environment of formation of these and more complex organics.

It is uncertain how much the original interstellar material has been altered in comets. There is some evidence of preservation from deuterium/hydrogen ratios in comets that are higher than nebular values (e.g., solar), and more representative of values in hot molecular cloud cores. But there is also evidence of chemical processing in comets.[19]

With the exception of in situ observations of organics in comet 1P/Halley, most of the organics known in comets have been detected as parent molecules in spectra obtained at millimeter, submillimeter, and infrared wavelengths. HCN was first detected in comet 1P/Halley in the radio region of the electromagnetic spectrum by several telescopes, and it was post-Halley (i.e., in the last 10 years) that other species were detected in comets. The field of comet parent-molecule radio spectroscopy has blossomed only recently, with the advent of new low-noise receivers. A summary of carbon-bearing species in cometary comas is shown in Table 4.1. While the approximate abundances are given for comets in general, these are based on the observations of only a few comets. In some cases, there are significant differences in the amount of organic compounds with different volatilities between comets, and this variability has implications concerning the location in the solar nebula where they were formed.[20]

Overall, it is estimated that perhaps as much as one-third of the comet nucleus by mass consists of organic solids, and carbon may be present at the 1 to 10 percent level in other volatiles such as CH_3OH, CO, CO_2, and others. Some of these molecules originate from the comet nucleus in the coma, but others have extended sources, originating presumably on the surfaces of the CHON grains. These include CO, H_2CO, OCS, HCN, and CS.

The neutral gas mass spectrometer aboard the European Space Agency's Giotto provided measurements of several native species, including the retrospective identification of ethane.[21] The so-called IKS infrared spectrometer on the former Soviet Union's Vega 1 spacecraft achieved detections of H_2O, CO_2, H_2CO, and possibly CO.[22] This instrument also detected a spectral signature in the 3.2- to 3.6-μm region attributed primarily to C-H stretching vibrations in one or more organics. The Particle Impact Analyser dust composition experiment on the Giotto spacecraft discovered a population of dust particles composed of an organic material containing CHON grains. These detections are included in Table 4.1. Most of the carbon found in comet 1P/Halley from mass spectrometry consisted of oxidized compounds, e.g., CO, CO_2, and H_2CO, and the refractory materials can be interpreted as being the following types of carbonaceous matter: pure carbon, polycyclic aromatic and highly branched aliphatic hydrocarbons, and polymers of carbon suboxide and of cyanopolyynes.[23,24] Many of the grains contained mixtures of complex organic compounds consistent with the presence of alcohols, aldehydes, ketones, acids, and amino acids. Other organic compounds included PAHs, highly branched aliphatic hydrocarbons, and unsaturated hydrocarbon chains, including alkynes, dienes, various large nitriles, and analogous species with imino and amino end groups. Several nitrogen heterocycles may also have been identified, including pyrrole, pyrroline, pyridine, pyrimidine, imidazole, purine and adenine.

The abundances listed in Table 4.1 are consistent with the abundances seen in interstellar ices or in hot molecular cloud cores where the molecules may have evaporated off ices. However, not all of the species listed are considered to be parent molecules (e.g., direct nucleus sublimation products), or unaltered molecules preserved from the interstellar medium. It is very important to note that the data in Table 4.1 are from a very small sample of comets: primarily from comets C/1996 B2 (Hyakutake) and C/1995 O1 (Hale-Bopp). The significance of these two objects is that they were the first really bright comets discovered far enough in advance to be observable over a range of distances following the development of adequate technology for the detection of parent molecules.

The study of cometary chemical composition has evolved rapidly in recent years, mainly owing to advances in instrumentation, especially high-dispersion cryogenic spectrometers at infrared wavelengths. High-resolution ($\lambda/\Delta\lambda > 10^4$) infrared spectroscopy enables individual rotational-vibrational lines to be resolved and differentiated from telluric absorptions (i.e., absorptions in Earth's atmosphere), other molecular emissions, and the continuum, making accessible species that are not observable with low resolution (Figure 4.2). The discovery of methane (CH_4), ethane (C_2H_6), and acetylene (C_2H_2) in comet C/1996 B2 (Hyakutake)[25] established the importance of these

TABLE 4.1 Carbon-Bearing Species Abundances in Comets Relative to Water

Observed				Inferred		
Molecule	Abundance	Technique		Molecule	Abundance	Technique
H_2O	100	UV, IR, radio		CH_3CHO	0.005	Mass spec[a]
CO	1-30	UV, IR, radio		C_3H_2	0.001	Mass spec[a]
CO_2	3-10	IR		C_2H_5CN	0.00028	Mass spec[a]
HCO		Radio		CH_3NH_2	<0.0015	Mass spec[a]
H_2CO	0.1-1.1	IR, radio				
CH_3OH	1-7	IR, radio				
HCOOH	0.08	Radio[b]				
HNCO	0.1	Radio[b]				
$HCONH_2$	0.01	Radio[b]				
HCO_2CH_3	0.08	Radio[b]				
CH_3CHO	0.02	Radio[b]				
CH_2	0.0027	Mass spec[a]				
CH_4	0.6	IR[b]				
C_2H_2	0.1	IR				
C_2H_4	0.003	Mass spec[a]				
C_2H_6	0.1-0.5	IR				
CH_3C_2H	<0.045	Radio[b]				
HCN	0.05-0.2	IR, radio				
HNC	0.04	Radio				
CH_3CN	0.02	Radio[b]				
HC_3N	0.03	Radio[b]				
NH_2CHO	0.01	Radio[b]				
CS	0.2	UV, radio				
CS_2	0.1-0.2	Radio[b]				
H_2CS	0.02	Radio[b]				
OCS	0.4-0.5	IR, radio[b]				
H_2CS	0.02	Radio[b]				
OCS	0.4-0.5	IR, radio[b]				
OCS	0.4-0.5	IR, radio[b]				
OCS	0.4-0.5	IR, radio[b]				

NOTE: This table does not include species for which only upper limits for nondetections are known.

[a]Comet Halley mass spectrometer data.

[b]Measurement from comet C/1995 O1 (Hale-Bopp) only.

SOURCE: Data from W.M. Irvine and E.A. Bergin, "Molecules in Comets: An ISM-Solar System Connection?," p. 447 in *Astrochemistry: From Molecular Clouds to Planetary Systems*, Proceedings of IAU Symposium 197 (Y.C. Minh and E.F. van Dishoeck, eds.), Publications of the Astronomical Society of the Pacific, San Francisco, Calif., 2000; K. Altwegg, H. Balsiger, and J. Geiss, "Composition of the Volatile Material in Halley's Coma from In Situ Measurements," *Space Science Reviews* 90: 3-18, 1999; D. Bockelée-Morvan, D.C. Lis, J.E. Wink, D. Despois, J. Crovisier, R. Bachiller, D.J. Benford, N. Biver, P. Colom, J.K. Davies, E. Gérard, B. Germain, M. Houde, D. Mehringer, R. Moreno, G. Paubert, T.G. Phillips, and H. Rauer, "New Molecules Found in Comet C/1995 O1 (Hale-Bopp): Investigating the Link Between Cometary and Interstellar Material," *Astronomy and Astrophysics* 353: 1101-1114, 2000; and P. Ehrenfreund, W. Irvine, L. Becker, J. Blank, J.R. Brucato, L. Colangeli, S. Derenne, D. Despois, A. Dutrey, H. Fraaije, A. Lazcano, T. Owen, and F. Robert, "Astrophysical and Astrochemical Insights into the Origin of Life," *Reports on Progress in Physics* 65: 1427-1487, 2002.

hydrocarbons in comets. Relative abundances provide crucial information about the link between interstellar and cometary ices. A larger-format high-efficiency array allowing simultaneous spectral order sampling from 1 to 5.5 μm became available as a facility instrument on the Keck II 10-m telescope in 1999 (NIRSPEC), and this capability is revolutionizing the characterization of the distribution of organics in small bodies. With only three grating settings, NIRSPEC permits a nearly complete high-dispersion survey of the spectral region from 2.9 to 3.7 μm, which is key for investigating the organic composition of primitive bodies. However, intense observing

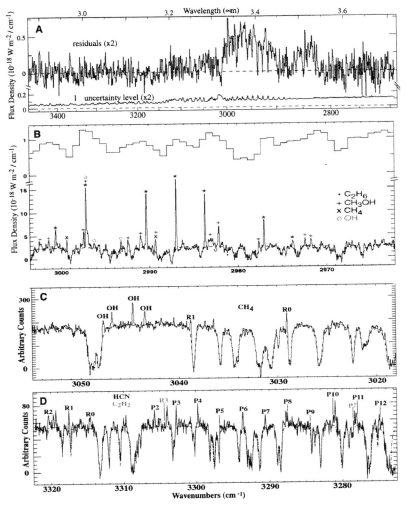

FIGURE 4.2 A comparison of high- and moderate-resolution cometary spectra. (A) Shown is an extracted spectrum of comet C/1999 H1 (Lee) taken in the moderate dispersion mode ($\lambda/\Delta\lambda \sim 2000$) on UT on August 20, 1999. The broad feature centered at 3.52 μm is due to CH_3OH ν_3 and shows emission from the P-, Q-, and R-branches. The broad feature between 3.3 and 3.45 μm is due mainly to unresolved organic emission lines. The channel-by-channel noise amplitude residuals (1σ) are also shown. (B) High-dispersion spectra ($\lambda/\Delta\lambda \sim 25,000$) of comet C/1999 H1 (Lee) acquired on UT August 21, 1999. Shown for comparison at the top of the panel is the moderate-resolution spectrum from (A) in this wavelength range. Ethane (C_2H_6), methanol (CH_3OH), methane (CH_4), and OH prompt emission, clearly detected in the high-dispersion spectrum ($\lambda/\Delta\lambda \sim 25,000$), are not resolved at $\lambda/\Delta\lambda \sim 2000$. (C) High-dispersion spectra ($\lambda/\Delta\lambda \sim 25,000$) of comet C/2000 WM1 (LINEAR) acquired on UT November 23, 2001. Methane and OH prompt emission were detected. (D) High-dispersion spectra ($\lambda/\Delta\lambda \sim 25,000$) of comet C/2000 WM1 (LINEAR) acquired on UT November 23, 2001. Rovibrational lines of hydrogen cyanide (HCN) and acetylene (C_2H_2) were detected. Symmetric hydrocarbons C_2H_2, C_2H_6, and CH_4 can be detected only at infrared wavelengths and were not detected until high-resolution spectroscopy became available. (See T.Y. Brooke, A.T. Tokunaga, H.A. Weaver, J. Crovisier, D. Bockelée-Morvan, and D. Crisp, "Detection of Acetylene in the Infrared Spectrum of Comet Hyakutake," *Nature* 383: 606-608, 1996; and M.J. Mumma, M.A. DiSanti, N. Dello Russo, M. Fomenkova, K. Magee-Sauer, C.D. Kaminski, and D.X. Xie, "Detection of Abundant Ethane and Methane, Along with Carbon Monoxide and Water, in Comet C/1996 B2 Hyakutake: Evidence for Interstellar Origin," *Science* 272: 1310-1314, 1996.) This illustrates the difficulty of detecting and isolating organic species with low or moderate spectral resolution, and the need for high-resolution spectroscopic studies to provide a more complete picture of cometary chemistry. SOURCE: Image courtesy Michael J. Mumma and Neil Dello Russo, NASA.

pressure for this limited resource has meant that progress in infrared spectroscopic detection of parent species has been slow.

A'Hearn and colleagues have undertaken an extensive study of abundances and molecular production rates for C_2, C_3, and several noncarbon-bearing species.[26] They found that most comets were similar in overall chemical composition. However, there was a group of comets depleted in the carbon-chain molecules. This depletion correlates strongly with dynamical age, which is related to location of formation in the solar nebula; i.e., only the Jupiter-family short-period comets show the depletion. A'Hearn and colleagues argue that this correlation is probably a consequence of primordial rather than evolutionary difference, with some process destroying carbon-chain molecules in the vicinity of the Kuiper belt (the source region of the Jupiter-family short-period comets).

Carbon Isotopes

Time-of-flight mass spectrometry for comet 1P/Halley found that the isotopic composition of carbon had an unusual signature. The $^{12}C/^{13}C$ ratios in some of the grains were up to 5000, suggesting that some of the grains may have originated from different interstellar environments (similar to what is observed for some chondrites).[27]

Delivery to Earth

Cometary impacts with Earth today are rare, but they may have been more important on primitive Earth when they were ejected into the Oort cloud in the first few 10^7 years of the solar system's history. While some researchers maintain that comets could have played an important role in the origins of life by delivering water and biologically relevant organic molecules to the primitive Earth, not all agree. Many researchers argue that the origin of life likely requires simple mixtures of a few relatively pure compounds and not the complex mixtures of highly processed carbon compounds likely to be found on comets.[28] Even if small amounts of biologically important prebiotic compounds are transported to Earth by comets, it is not known if these organics survived the collision with Earth's atmosphere. Fullerenes, formed in the interstellar medium, were detected in the Sudbury impact site in Canada.[29] While some researchers argue that the Sudbury structure is the result of a cometary impact,[30] the detection of fullerenes remains controversial. But, in addition to not being a good starting point for the origins of life, fullerenes, polycyclic aromatic hydrocarbons, and other complex organic molecules are not even good foodstuffs for biochemistry. Indeed, the conversion of cometary carbon compounds into biomass may require their oxidation to carbon dioxide and subsequent fixation by preexisting biochemical processes.

Centaurs and Kuiper Belt Objects

Kuiper belt objects and Centaurs (with few exceptions) never get close enough to the Sun to be active and bright enough for direct detection of carbon-bearing species. The presence of organics must be inferred from solid-state infrared spectra of the nuclei or from the colors of the surface materials. Measurements of the thermal emission from some of the Centaurs show that they have generally low albedos.

Centaurs and Kuiper belt objects display a range of surface colors from neutral (solar) to very red for nearly 150 objects that have been measured. Near-infrared spectroscopy of several dozen Centaurs and KBOs has revealed a similarly large diversity. Water-ice absorption features are present on a few, and spectral models suggest surfaces probably rich in complex organic solid materials such as amorphous carbon, methanol ice, other light hydrocarbons, and Titan and Triton tholins[31] in the presence of olivine. Table 4.2 lists the Centaurs and KBOs for which spectra exist, along with the possible identification of the surface compositions as derived from spectral models.

This diversity seen in the surface properties is likely to be the result of several processes. One possibility is a competition between a reddening of the surface by high-energy particles and collisions that can excavate more neutral material. However, statistical work on the distribution of the colors has led to contradictions with this model. For instance, the dispersion of colors over all the objects is much larger than the dispersion of colors for any given object (over rotational phase), suggesting that the bodies have uniform colors, whereas the collision/

TABLE 4.2 Carbon Compounds Inferred on Surfaces of Centaurs and Kuiper Belt Objects

Object	Spectral Features	Type[a]	Object	Spectral Features	Type[a]
(15789) 1993 SC	Hydrocarbons	KBO	(15874) 1996 TL$_{66}$	Featureless	KBOS
(26181) 1996 GQ$_{21}$	Flat, Titan tholin, H$_2$O, C	KBOS	1996 TS$_{66}$	Featureless	KBO
(19308) 1996 TO$_{66}$	Strong H$_2$O	KBO	(28978) 2001 KX$_{76}$	Featureless	KBO
(47171) 1999 TC$_{36}$	Titan tholin, amorphous C, H$_2$O	KBO	(2060) Chiron	Featureless; H$_2$O	Cent
(26375) 1999 DE$_9$	H$_2$O, organic, C	KBOS	5145 Pholus	Titan tholins, olivine, H$_2$O, CH$_3$OH, amorphous C	Cent
(38628) 2000 EB$_{173}$	Featureless, H$_2$O?	KBO	(52872) Okyrhoe	H$_2$O, kerogen, olivine	Cent
(20000) 2000 Varuna	H$_2$O?	KBO	(32532) PT$_{13}$	Titan tholin, ice tholin, olivine, amorphous C	Cent
(54598) 2000 QC$_{243}$	H$_2$O, kerogen, olivine	Cent	(8405) Asbolus	Featureless; Triton tholin, Titan tholin, amorphous C, ice tholin	Cent
(63252) 2001 BL$_{41}$	Triton tholin, ice tholin, amorphous C	Cent	(10199) Chariklo	Titan, Triton tholins, amorphous C, H$_2$O	Cent
(31824) 1999 UG$_{35}$	H$_2$O, amorphous C, Titan tholin, CH$_3$OH, olivine	Cent	1996 TQ$_{66}$	1.9-μm feature— identity unknown	KBO
1996 TS$_{66}$	1.9-μm feature— identity unknown	KBO			

NOTE: Review papers that summarize the near-infrared spectra and spectral model results are quoted in K.J. Meech, O.R. Hainaut, H. Boehnhardt, and A. Delsanti, "Search for Cometary Activity in KBO (24952) 1997 QJ$_4$," *Earth, Moon, and Planets* 92: 169-181, 2003; and E. Dotto, M.A. Barucci, and C. de Bergh, "Colours and Composition of the Centaurs," *Earth, Moon, and Planets* 92: 157-167, 2003.

[a] KBO, low-eccentricity Kuiper belt object; KBOS, high-eccentricity KBO; Cent, Centaur.

reddening model predicts variegated surfaces. Another suggestion, based on the observation of a color trend with inclination,[32] is that the high-inclination objects form a dynamically "hot" (excited) population of objects that were scattered during planet migration from smaller heliocentric distances. In this scenario, the color/compositional diversity might simply reflect different source regions.

Cometary-like volatile activity could cause a change in surface albedo by removing some of the older, reddened surface and replacing it with bluer material. This hypothesis has been suggested as the mechanism to explain the change in the color and rotational light curve for object (19308) 1996 TO$_{66}$.[33]

Trojan Asteroids, Small Outer Solar System Satellites, and Rings

Small Jovian Satellites and Trojans

There is only sparse multispectral data for the jovian ring and inner satellites, with observations from the Hubble Space Telescope's Near-Infrared Camera and Multi-Object Spectrometer (NICMOS) showing similar positive spectral slopes through the near-infrared, suggesting similar dusty compositions possibly due to contamination by Io. The minor outer satellites (likely captured asteroids) of Jupiter are spectrally dark and neutral to red in the visible, although they lack the ultra-red material seen on some of the Centaurs. In the near-infrared, the retrograde satellites show more spectral heterogeneity than do the prograde satellites, and have spectra and presumably compositions similar to D-type asteroids, whereas the prograde satellites are compositionally more similar to C-type asteroids. The Trojan asteroids have low mean geometric albedos and flat optical reflection spectra with a range of spectral slopes from neutral to reddish, implying surfaces rich in complex organic material. Their spectra are very similar to those of comet nuclei and to that of the Tagish Lake meteorite.[34] A summary of the outer solar system satellite carbon-bearing materials is given in Table 4.3.

Phoebe, Hyperion, Iapetus, and Enceladus

The outermost of Saturn's major satellites, Phoebe, has a relatively flat spectrum thought to be composed of carbonaceous chondritic material. Hyperion shares spectral properties with the uranian satellites. The best laboratory fit to the spectrum of Hyperion obtained with Earth-based telescopes comes from a mixture of water ice and spectrally neutral charcoal. The latter suggests the presence of dark, carbon-bearing materials.

Iapetus has the most striking large-scale albedo contrast of any solar system body. While its trailing hemisphere has an albedo of ~0.5 and is composed of nearly pure ice, material in its leading hemisphere has an albedo of <0.05 with an especially steep red spectral slope through the visible. The spectrum has been modeled with Murchison organic residue, polymeric HCN, and water ice.[35] Visible and near-infrared spectra suggest a spectral match to the organic fraction of carbonaceous chondrites along with iron-rich hydrated minerals, i.e., clays. The source of the dark material being exogenic (e.g., impacted dust) or endogenic (perhaps concentrated in the leading hemisphere by impacts) is debated.

Observations of Enceladus from the Cassini spacecraft have recently revealed the presence of an atmospheric plume and coma dominated by water, with significant amounts of carbon dioxide, methane, and other constituents, possibly carbon monoxide or molecular nitrogen. Trace quantities (<1 percent) of acetylene and propane also appear to be present.[36] The plume, or rather plumes, of material appears to be venting from a series of anomalously

TABLE 4.3 Carbon Compounds Inferred on Surfaces of Small Outer Solar System Bodies

Planet System	Object	Carbon Inferred	References
Jupiter	Prograde outer satellites	C-type asteroidal material	a
	Retrograde outer satellites	D-type asteroidal material	a
	Trojan asteroids	P- and D-type asteroidal material	b
Saturn	Hyperion	D- or T-type asteroidal material	c,d
	Iapetus	Organic-rich, hydrated silicates	e,f
	Phoebe	C-type asteroidal material	c
Uranus	Minor inner satellites	Dark material	g
	Rings	Dark spectrally neutral material	h
Neptune	Minor satellites and rings	Dark spectrally neutral material	i

[a]M.V. Sykes, B. Nelson, R.M. Cutri, D.J. Kirkpatrick, R. Hurt, and M.F. Skrutskie, "Near-infrared Observations of the Outer Jovian Satellites," *Icarus* 143: 371-375, 2000.

[b]J.C. Gradie, C.R. Chapman, and E.F. Tedesco, "Distribution of Taxonomic Classes and the Compositional Structure of the Asteroid Belt," Pp. 316-335 in *Asteroids II* (Richard P. Binzel, T. Gehrels, and Mildred Shapley Matthews, eds.), University of Arizona Press, Tucson, Ariz., 1989.

[c]D. Tholen and B. Zellner, "Eight-color Photometry of Hyperion, Iapetus, and Phoebe," *Icarus* 53: 341-347, 1983.

[d]K.S. Jarvis, F. Vilas, S.M. Larson, and M.J. Gaffey, "Are Hyperion and Phoebe Linked to Iapetus?" *Icarus* 146: 125-132, 2000.

[e]J.F. Bell, D.P. Cruikshank, and M.J. Gaffey, "The Composition and Origin of the Iapetus Dark Material," *Icarus* 61: 192-207, 1985.

[f]F. Vilas, S.M. Larson, K.R. Stockstill, and M.J. Gaffey, "Unraveling the Zebra: Clues to the Iapetus Dark Material Composition," *Icarus* 124: 262-267, 1996.

[g]P. Thomas, C. Weitz, J. Veverka, "Small Satellites of Uranus—Disk-integrated Photometry and Estimated Radii," *Icarus* 81: 92-101, 1989.

[h]C.C. Porco, J.N. Cuzzi, M.E. Ockert, and R.J. Terrile, "The Color of the Uranuan Rings," *Icarus* 72: 69-78, 1987.

[i]B.A. Smith, L.A. Soderblom, D. Banfield, C. Barnet, R.F. Beebe, A.T. Bazilevskii, K. Bollinger, J.M. Boyce, G.A. Briggs, and A. Brahic, "Voyager 2 at Neptune—Imaging Science Results," *Science* 246: 1422-1449, 1989.

warm, trenchlike rifts in Enceladus's southern polar region. Additional remote sensing and in situ observations of the venting regions, the plumes, and their organic content will be a high priority for the remainder of Cassini's prime mission and will likely be an even higher priority during an extended mission.

The Uranian and Neptunian Systems

The uranian satellites are notably dark. The five major satellites (Miranda, Ariel, Umbriel, Titania, and Oberon) have average normal reflectances ranging from a high of ~33 percent for Miranda down to ~19 percent for Oberon and Umbriel. There is a general anticorrelation of visible geological activity and albedo on their surfaces, with the overall brightest (Miranda) showing the most geological activity. The floor of a large basin on Oberon shows the darkest material observed on these five satellites, with normal reflectance of ~10 percent. Some of darkest materials on Miranda are along prominent scarps interpreted as normal faults, perhaps exposing a non-uniform subsurface layer of dark material. The satellites' Earth-based telescopic spectra have been modeled with mixtures of spectrally neutral charcoal and water ice, and possibly ammonium hydrate for Miranda.[37] Voyager observations show that they are somewhat red, notably on the leading hemispheres of the outer four, suggesting accumulation of meteoritic dust.

The small satellite Puck, and potentially the other small inner uranian satellites, have albedos of ~0.07, but their spectral characteristics are uncertain. The narrow uranian rings appear to be spectrally neutral and dark, suggesting carbon as an important component. The spectroscopic and geological evidence leaves open the question of whether dark surfaces in the uranian system include primordial silicate and/or carbon-bearing dark materials, or have darkened in situ due to energetic processing and polymerization of carbon-bearing ices.

Little is known of the composition of Neptune's satellite Nereid and its six small inner satellites. These satellites have relatively flat visible spectra and low albedos.

SUMMARY OF PAST, PRESENT, AND PLANNED MISSIONS: IMPLICATIONS FOR CARBON STUDIES

Comets have been proposed as a potential source of organics that were delivered to the inner planets. As noted previously, the organics in comets are probably more closely related structurally to the interstellar molecules than any other organics in the solar system. Currently three missions to primitive bodies are in progress or under development, and six have successfully completed their primary missions:

- *Galileo.* The Galileo spacecraft flew by asteroid 951 Gaspra on October 29, 1991, and by 243 Ida and its moon, Dactyl, on August 28, 1993. Near-Infrared Mapping Spectrometer (NIMS) data in the range from 0.7 to 5.3 µm showed that Gaspra is olivine-rich with a high olivine/pyroxene abundance ratio, and that Ida has an orthopyroxene/(orthopyroxene + olivine) ratio consistent with LL chondrites. Both are S-type asteroids. The spectrum of Dactyl also features a relatively deep 0.97-µm absorption.[38] No organic features were found.
- *NEAR.* The Near-Earth Asteroid Rendezvous (NEAR) spacecraft was launched on February 17, 1996, and carried an instrument package optimized to examine asteroid mineralogy and elemental abundances. Its instruments included an x-/gamma-ray spectrometer, a multispectral imager, and a near-infrared spectrometer that could provide spatially resolved spectra between 0.8 and 2.5 µm. On June 27, 1997, NEAR flew by the C-type asteroid 253 Mathilde and obtained spatially resolved spectra of one side of this 66 km by 48 km by 46 km body. The color of Mathilde is consistent with that of CM carbonaceous chondrites.

  On February 14, 2000, NEAR entered orbit about the S-type asteroid 433 Eros. For the next year it conducted a global survey of this 34 km by 11 km by 11 km body before finally landing on its surface on February 12, 2001. Resolved spectra showed mineral absorption features at 1 and 2 µm that were characteristic of orthopyroxene and olivine, and spectral models suggested that the red color of the surface was due to space weathering of nanophase iron.[39] The infrared spectral data and results of the x-ray spectrometer suggested a composition similar to that of a relatively primitive ordinary chondrite.[40,41]

- *Deep Space 1.* The Deep Space 1 mission was a technology validation mission that was launched in October 1998 carrying a camera and ultraviolet (0.08 to 0.18 μm) and infrared (1.2 to 2.4 μm) spectrophotometers. The main purpose of the mission was to test a solar-electric propulsion system. The spacecraft had a close encounter with asteroid 9969 Braille on July 28, 1999. Initial indications were that the spectrum of Braille was similar to that of the asteroid Vesta. This interpretation proved untenable because the center of the 1-μm absorption band is at much too long a wavelength, and it is now accepted that the spectrum of Braille is a good match to that of a Q-type asteroid.[42] Subsequent studies by the Deep Space 1 science team have confirmed this classification and have also suggested that Braille's spectrum closely matches that of the most commonly found meteorites, the ordinary chondrites.[43] Following the encounter, the spacecraft then embarked on its extended science mission to comet 19P/Borrelly (see Figure 4.3). The first in situ images were obtained of a Jupiter-family short-period comet and its surrounding environment on September 21, 2001, revealing a low-albedo (i.e., very black) nucleus.[44] The mission ended in December 2001.
- *Stardust.* The Stardust spacecraft was launched by NASA in 1999 and successful encountered the nucleus of the short-period comet 81P/Wild 2 (see Figure 4.3) on January 2, 2004. In situ images of the nucleus at closest approach showed a low-albedo nucleus covered in crater-like features. The spacecraft captured particles coming from the coma of the comet and successfully returned them to Earth for analysis on January 15, 2006. In addition to the aerogel used to capture the dust grains, the spacecraft carried a Cometary and Interstellar Dust Analyzer (CIDA), a dust flux monitor, and a navigation camera. CIDA, a mass spectrometer, provided some information about possible organic compounds. It is estimated that several thousand individual particles were collected on the back-to-back aerogel collecting trays as the spacecraft passed through the coma. The collectors were withdrawn into a reentry capsule that parachuted to Earth. The particles were then transferred to a laboratory for analysis. The samples will ultimately provide data on the amounts and classes of organics present in comets. To date, the most perplexing result from the initial analysis of the cometary samples is the abundance of mineral grains that could only have originated in the hottest, innermost regions of the solar nebula. Meanwhile, the Stardust spacecraft (minus its sample-return canister) continues to operate, and it may be redirected to encounter additional comets or asteroids.
- *Hayabusa.* The Hayabusa mission, formerly known as MUSES-C, is a Japanese undertaking to rendezvous with 25143 Itokawa (a small near-Earth S-type asteroid, with a semimajor axis of 1.32 AU) and collect a surface sample for return to Earth. The spacecraft was launched on May 9, 2003, performed an Earth gravity-assist maneuver in May 2004, and used its solar-electric propulsion system to match orbits with the ~700 m by 300 m by 300 m body on September 12, 2005. Hayabusa's instrumentation consists of a LIDAR (operating at 1.064 μm and used for ranging, gravitational field and surface reflectivity measurements, and topographic mapping), an x-ray fluorescence spectrometer (to determine the global abundances of iron, solium, magnesium, aluminum, silicon, sulfur, calcium, and titanium on the asteroid's surface), a multiband imager (to enable topographic mapping and multiband polarimetry in eight spectral bands between 0.30 and 1.10 μm), and a near-infrared spectrometer (to obtain high-resolution mineralogical maps at wavelengths from 0.85 to 2.10 μm).

These instruments were used successfully to conduct a global survey of the asteroid from an altitude as low as 10 km. But then things began to go wrong. The attempt to deploy Minerva, a 0.55-kg lander equipped with three camera systems and temperature sensors, failed. Instead of using a pair of torquers to enable it to hop across the surface of Itokawa, Minerva drifted off into space and was lost. On November 19, 2005, and again 6 days later, Hayabusa attempted to perform touch-and-go landing at different sites on the asteroid's surface.[45] During each momentary touch down, the spacecraft was supposed to fired a 5-gram tantalum projectile into Itokawa's surface. The debris thrown up from the surface was to be directed into a sample-return canister via a 1-m-long aluminum and fabric collecting funnel. It is not clear if any of these actions actually occurred. Multiple problems appear to have arisen during both attempts to conduct the complex, autonomously directed sample-collection process. These problems, combined with the failure of other spacecraft subsystems, ultimately delayed Hayabusa's departure from Itokawa, and the return of the samples to Earth, if any were actually collected. It is now expected to reach Earth in 2010. In the

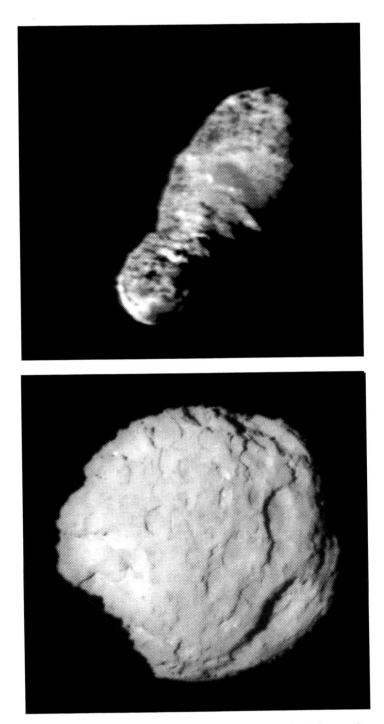

FIGURE 4.3 (Top): Image of the nucleus 19P/Borrelly taken by the Deep Space 1 mission on September 25, 2001, from a distance of 3417 km. The nucleus is 8 km long (nuclei range in size between 0.1 and 100 km). The albedo varies between 0.007 and 0.03 across the surface. (Bottom): The closest short exposure of the comet Wild 2, which NASA's Stardust spacecraft flew by on January 2, 2004, taken at an 11.4-degree angle, the angle between the camera, comet, and the Sun. SOURCE: Images courtesy of NASA/JPL.

meantime, the Japan Aerospace Exploration Agency is investigating the possibility of building an improved version of Hayabusa and sending it to another asteroid sometime in the future.

- **Deep Impact.** The main goal of the Deep Impact mission was to examine the pristine interior of a comet by creating a controlled experiment to excavate beneath the evolved surface layers. The Deep Impact spacecraft was launched on January 12, 2005, and released a 360-kg impactor on July 2, 2005. Two days later, while the Deep Impact spacecraft remained at a safe distance, the impactor collided with the nucleus of the Jupiter-family comet 9P/Tempel 1. The impact was observed by an imager on the impactor, as well as by the medium- and high-resolution optical cameras and near-infrared spectrometer (1.0 to 4.8 μm) on Deep Impact. The spectroscopic capabilities were designed to cover the organically interesting region of the near-infrared spectrum. The team was particularly interested in measuring the abundances of H_2O, H_2CO, CO_2, CO, and other organic species before, during, and after impact (Figure 4.4). Deep Impact continues to operate and may be redirected to encounter additional objects in the future.
- **Dawn.** The goal of the solar-electric-powered Dawn mission, currently scheduled for launch in June 2007, is to sequentially orbit the large, main-belt asteroids 1 Ceres and 4 Vesta to study their basic structures and compositions. Despite numerous technical and programmatic issues during its development, including cancellation in February 2006 and reinstatement a few months later, Dawn is still scheduled to arrive at Vesta in October 2011 and at Ceres in February 2015. The Dawn mission instrumentation includes an imaging camera, gamma-ray and neutron spectrometer, and a visible and infrared mapping spectrometer (0.35 to 0.9 μm, 0.8 to 2.5 μm, and 2.4 to 5.0 μm, respectively). One of the top-level questions addressed will be to assess the role of size and water in the evolution of planetary bodies. The 960-km-diameter Ceres is a low-albedo G-type asteroid, which exhibits an infrared reflectance spectrum possibly associated with the effects of aqueous alteration. The 520-km-diameter Vesta is a V-type asteroid and likely source of the so-called Howardite, Eucrite, and Diogenite (HED) meteorites.[46]
- **Rosetta.** The European Space Agency launched Rosetta to comet 67P/Churyumov-Gerisimenko in March 2004, and it is expected to arrive at its target in November 2014. The mission goals are to undertake a long-

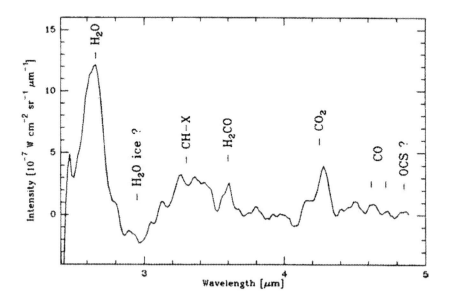

FIGURE 4.4 Infrared spectrum of comet 1P/Halley showing emission lines in the region of the spectrum of interest for Deep Impact. SOURCE: Reprinted from M. Combes, J. Crovisier, T. Encrenaz, V.I. Moroz, and J.-P. Bibring, "The 2.5-12 Micron Spectrum of Comet Halley from the IKS-VEGA Experiment," *Icarus* 76: 404-436, copyright 1988, with permission from Elsevier.

term in situ exploration of the comet from large heliocentric distances through perihelion. The scientific payload of Rosetta includes a visual infrared spectral and thermal spectrometer (VIRTIS), which will be fundamental for characterizing specific spectral bands of minerals and organic molecules arising from surface components and from materials dispersed in the coma. Their identification is a primary goal of the Rosetta mission.[47] In addition, the orbiter mission instrumentation includes an ultraviolet spectrometer, a nucleus sounder, an ion mass analyzer, a grain impact analyzer, a microwave system, an imaging system, a plasma experiment, and a radio science experiment. The lander will also be well instrumented, including the cometary sampling and composition (COSAC) experiment and an evolved gas analyzer (PTOLEMY). COSAC is designed to identify complex organic molecules, and PTOLEMY will make accurate measurements of the isotopic ratios of light elements. The team is also planning to fly by 1 to 2 asteroids en route (the targets are still to be identified).

- *New Horizons.* The first of the New Frontiers mission line, and the NRC solar system exploration decadal survey's highest recommendation for a new medium-class mission,[48] New Horizons is designed to characterize the primitive icy bodies in the far reaches of the solar system. The spacecraft was launched on schedule on January 19, 2006, and will reach the Pluto-Charon system in July 2015. After a flyby reconnaissance of the Pluto-Charon system, the spacecraft will continue outward to examine one or two KBOs. The primary objectives include characterization of the global geology and surface composition (H_2O, CO, CO_2, and CH_4) of Pluto and Charon, and the neutral atmosphere of Pluto. The primary instrument payload includes an ultraviolet/visible/infrared imaging and spectroscopy system, and a radio science/radiometry package. Low-resolution infrared spectroscopy is available between 1.25 and 2.50 μm.

FUTURE RESEARCH DIRECTIONS

Although there clearly is a wealth of recent, ongoing, and upcoming missions to primitive bodies, they have not all been optimized for the study of organics. Within the primitive small-body population there exists a preserved reservoir of organic material and some small bodies are trapped in dynamically stable niches spread over a wide range of nebular heliocentric distances. A rich research opportunity exists to explore these different chemical and thermal regimes, thus enabling an understanding of the distribution and history of primordial carbon in the solar system. Support for key existing and new capabilities in several areas related to research on small bodies can have a large impact toward furthering knowledge of organic matter in the solar system.

Laboratory Opportunities

The spectrally red colors of these primitive solar system bodies are often interpreted to be caused by the presence of organic materials. However, the spectral reflectance of Hektor, a D-type Trojan asteroid, can be modeled with a magnesium-rich pyroxene and a serpentine, and organic solids are not required to match the low albedo and red color of this object. Spectral models can account for the full range of colors of KBO and Centaur reflectance spectra as well as their low albedos from combinations of tholins and amorphous carbon.[49] These models also allow for (but do not require) the presence of space-weathered (reddened) igneous rock-forming minerals, and can incorporate the presence of volatiles in sufficient quantities that the absorption bands are not completely quenched by the coexistence of other low-albedo materials. Caution should be used within this type of model in inferring the organic inventory of these bodies. Observed changes in spectral slopes or broad, weak bumps or hollows can have no analytical relevance because color and albedo are highly nonlinear functions of composition and grain sizes and shapes. Thus the surface composition of mixtures (geometric and intimate) can be strongly influenced by the details of how the components are mixed. However, by adding information about the albedo of the object, many of the ambiguities can be reduced.[50] The Spitzer Space Telescope will undertake to measure the albedos of a large number of these outer solar system bodies, and thus researchers will soon be extremely well situated to investigate the organic compositions of these bodies with a combination of infrared spectroscopy and spectral modeling.

Additional laboratory spectral measurements of model compounds and materials are needed to provide further information concerning organic materials possibly found on Triton, Pluto, Charon, Centaurs, and Kuiper belt objects. The spectral properties of the ices of relevant hydrocarbons should also be studied in order to facilitate the detection of the various organics that may exist on these bodies. These laboratory-based investigations should utilize mixtures of candidate organic compounds alone and together with water ice to examine effects of mixing. Effects of irradiation of these ices should be examined to determine what organic compounds are formed.

At present, the relevant optical constants have been measured for only a few of the organic and inorganic compounds that are likely to be present in primitive bodies of interest. Without a suite of materials with known constants to incorporate in the spectral models, the identification of many of the observed spectral features remains challenging. With modest support for laboratory work of this kind, great progress could be made in understanding the organic components of primitive solar system bodies.

Recommendation: The physical, chemical, and spectroscopic properties of ices of potential hydrocarbon species should be studied to facilitate the detection of organic materials.

Ground-Based Observing Facilities and Opportunities

High-resolution infrared spectroscopy of comets is playing an increasing role in characterizing the link between cometary and interstellar ices and is providing insights into the chemistry of the protoplanetary nebula. While the science is exciting, progress in this area is likely to be slow because of the limited access to ground-based facilities capable of performing such observations (at present Keck with NIRSPEC, and Subaru with IRCS). Having more planetary science access to these facilities (Keck in particular), or a larger, modern replacement for the NASA Infrared Telescope Facility, would significantly increase the rate of progress in studies of the organic inventory in the solar system's most primitive objects. In addition, information is just beginning to accumulate about the organics on the Kuiper belt objects. The limit on gathering such information is access to large facilities capable of performing the low-resolution near-infrared spectroscopy of the solid surfaces (combined with spectral modeling).

Recommendation: The task group reiterates the call made in the 2003 report of the National Research Council's Solar System Exploration Decadal Survey Committee, *New Frontiers in the Solar System,*[51] that NASA's support for planetary observations with ground-based astronomical instruments, such as the Infrared Telescope Facility and the Keck telescopes, be continued and upgraded as appropriate, for as long as they provide significant scientific return and/or mission-critical support services.

Space Missions

The current and upcoming space missions are targeted to active comets, Pluto/Charon, and perhaps one or two KBOs, as well as to asteroids with a variety of spectral types. Many of the missions have at least limited capability to search for organics and carbon-bearing compounds, but only Deep Impact and Rosetta can obtain spectra out to 5 μm, and only Rosetta is truly optimized for organic studies.

Recommendation: Every opportunity should be taken to direct space missions to small bodies to do infrared spectral studies of these targets, especially a D- or P-type asteroid, to determine if these dark bodies contain an appreciable amount of carbon compounds and, if so, whether they are the sources of the carbonaceous meteorites and dust reaching Earth.

In this regard, a possible opportunity to conduct such studies is as an adjunct to the Trojan Asteroid/Centaur Reconnaissance flyby mission described in the solar system exploration decadal survey. Although this mission was not ranked in the survey's final list of priorities, the possibility of using a single spacecraft to make a sequential flyby of three different classes of primitive bodies—i.e., a D- or P-type main-belt asteroid, a jovian

Trojan asteroid, and a Centaur—has sufficient merit to warrant additional study for possible implementation as a New Frontiers mission at some time in the future.

The successful landing of the NEAR spacecraft on the asteroid Eros demonstrated the feasibility of sending a probe to land on an asteroid. In situ analyses as well as sample-return missions should be performed for both asteroids and comets. Such missions will provide direct information on the nature of the organics present in asteroids and comets and information about whether or not the asteroids are the sources of meteorites and dust reaching Earth. They will also supply data on the structural information on the compounds coming from asteroids as compared to the organics derived from comets. If there are major structural differences between the cometary and asteroidal organics, that information will provide insight into the chemical reactions that took place in forming the organics in each reservoir. These data will also provide insight into the differences in conditions present where the cometary and asteroidal organics were formed. It may be possible to suggest what organic structures in comets or asteroids were the more likely contributors of chemical precursors of life to the early Earth. At present there is no consensus as to which starting materials initiated the origin of life or even if extraterrestrial organic compounds even played a significant role in life's origin.

Recommendation: In situ analyses as well as sample-return missions should be performed for both asteroids and comets. The task group points to the solar system exploration decadal survey report's recommendation for a New Frontiers-class Comet Surface Sample Return mission as an example of an activity that would greatly enhance understanding of the organic constituents of the solar system's primitive bodies.[52]

Not all existing and planned missions to primitive bodies have been optimized for the study of organic materials. The population of primitive small bodies, however, may preserve organic materials from a wide range of nebular heliocentric distances. A rich research opportunity exists to explore these different chemical and thermal regimes, thus enabling an understanding of the distribution and history of organic materials in the solar system.

NOTES

1. Centaurs are minor solar system bodies whose heliocentric orbit lies between Jupiter and Neptune and typically crosses the orbits of one of the other outer giant planets (Saturn, Uranus, Neptune). The orbits of the Centaurs are dynamically unstable due to interactions with the giant planets and are probably dynamically evolving from the Kuiper belt into short-period comet orbits.

2. The Kuiper belt is a region at a heliocentric distance of approximately 30 to 50 AU in the solar system populated by small (up to a few times 100-km diameter) bodies believed to be remnants from the formation of the solar system. This region is believed to be the source of some of the short-period comets.

3. A. Carusi, L. Kresak, E. Perozzi, and G.B. Valsecchi, *Long-term Evolution of Short-period Comets*, Adam Hilger, Ltd., Bristol, England, 1985.

4. D. Laufer, E. Kochavi, and A. Bar-Nun "Structure and Dynamics of Amorphous Water Ice," *Physical Review B* 36: 9219-9227, 1987.

5. R.E. Johnson, "Irradiation Effects in a Comet's Outer Layers," *Journal of Geophysical Research* 96: 17553-17557, 1991.

6. G. Strazzulla and R.E. Johnson, "Irradiation Effects on Comets and Cometary Debris," pp. 243-275 in *Comets in the Post-Halley Era* (R.L. Newburn, M. Neugebauer, and Jürgen H. Rahe, eds.), Kluwer Academic Publishers, Dordrecht, The Netherlands, 1991.

7. M.H. Moore, R.L. Hudson, and R.F. Ferrante, "Radiation Products in Processed Ices Relevant to Edgeworth-Kuiper Belt Objects," *Earth, Moon, and Planets* 92: 291-306, 2003.

8. D. Prialnik, A. Bar-Nun, and M. Podolak, "Radiogenic Heating of Comets by ^{26}Al and Implications for their Time of Formation," *Astrophysical Journal* 319: 993-1002, 1987.

9. M. Podolak and D. Prialnik, "Conditions for the Production of Liquid Water in Comet Nuclei," pp. 231-234 in *A New Era in Bioastronomy* (G. Lemarchand and K. Meech, eds.), ASP Conference Series, Vol. 213, 2000.

10. M.F. A'Hearn, R.L. Millis, D.G. Schleicher, D.J. Osip, and P.V. Birch, "The Ensemble Properties of Comets: Results from Narrowband Photometry of 85 Comets, 1976-1992," *Icarus* 118: 223-270, 1995.

11. The color of an object refers to the variation of reflectivity as a function of wavelength, and those objects that have increasing reflectance at longer wavelengths are termed "red."

12. M.A. Barucci, D.P. Cruikshank, S. Mottola, and M. Lazzarin, "Physical Properties of Trojan and Centaur Asteroids," pp. 273-287 in *Asteroids III* (W.F. Bottke, Jr., A. Cellino, P. Paolicchi, and R.P. Binzel, eds.), University of Arizona Press, Tucson, Ariz., 2002.

13. T. Hiroi, M.E. Zolensky, and C.M. Pieters, "The Tagish Lake Meteorite: A Possible Sample from a D-Type Asteroid," *Science* 293: 2234-2236, 2001.

14. D.P. Cruikshank and C.M. Dalle Ore, "Spectral Models of Kuiper Belt Objects and Centaurs," *Earth, Moon, and Planets* 92: 313-330, 2003.

15. F.L. Whipple, "A Comet Model. I. The Acceleration of Comet Encke," *Astrophysical Journal* 111: 375-394, 1950.

16. M.J. Mumma., P.R. Weissman, and S.A. Stern, "Comets and the Origin of the Solar System—Reading the Rosetta Stone," pp. 1177-1252 in *Protostars and Planets III* (E. Levy and J.I. Lunine, eds.), University of Arizona Press, Tucson, Ariz., 1993.

17. W.M. Irvine, F.P. Schloerb, J. Crovisier, B. Fegley, Jr., and M.J. Mumma, "Comets: A Link Between Interstellar and Nebular Chemistry," p. 1159 in *Protostars and Planets IV* (V. Mannings, A.P. Boss, and S.S. Russell, eds.), University of Arizona Press, Tucson, Ariz., 2000.

18. J.M. Greenberg, "Making a Comet Nucleus," *Astronomy and Astrophysics* 330: 375-380, 1998.

19. M.J. Mumma, N. Dello Russo, M.A. DiSanti, K. Magee-Sauer, R.E. Novak, S. Brittain, T. Rettig, I.S. McLean, D.C. Reuter, and Li-H. Xu, "Organic Composition of C/1999 S4 (LINEAR): A Comet Formed Near Jupiter?" *Science* 292: 1334-1339, 2001.

20. N. Biver, D. Bockelée-Morvan, P. Colom, J. Crovisier, B. Germain, E. Lellouch, J.K. Davies, W.R.F. Dent, R. Moreno, G. Paubert, J. Wink, D. Despois, D.C. Lis, D. Mehringer, D. Benford, M. Gardner, T.G. Phillips, M. Gunnarsson, H. Rickman, A. Winnberg, P. Bergman, L.E.B. Johansson, and H. Rauer, "Long-term Evolution of the Outgassing of Comet Hale-Bopp from Radio Observations," *Earth, Moon, and Planets* 78: 5-11, 1999.

21. P. Eberhardt, "Comet Halley's Gas Composition and Extended Sources: Results from the Neutral Mass Spectrometer on Giotto," *Space Science Reviews* 90: 45-52, 1999.

22. M.J. Mumma and D.C. Reuter, "On the Identification of Formaldehyde in Halley's Comet," *Astrophysical Journal* 344: 940-948, 1989.

23. K. Altwegg, H. Balsiger, and J. Geiss, "Composition of the Volatile Material in Halley's Coma from In Situ Measurements," *Space Science Reviews* 90: 3-18, 1999.

24. M.N. Fomenkova, "On the Organic Refractory Component of Cometary Dust," *Space Science Reviews* 90: 109-114, 1999.

25. M.J. Mumma, M.A. Disanti, N. dello Russo, M. Fomenkova, K. Magee-Sauer, C.D. Kaminski, and D.X. Xie, "Detection of Abundant Ethane and Methane, Along with Carbon Monoxide and Water, in Comet C/1996 B2 Hyakutake: Evidence for Interstellar Origin," *Science* 272: 1310-1314, 1996.

26. M.F. A'Hearn, R.L. Millis, D.G. Schleicher, D.J. Osip, and P.V. Birch, "The Ensemble Properties of Comets: Results from Narrowband Photometry of 85 Comets, 1976-1992," *Icarus* 118: 223-270, 1995.

27. E.K. Jessberger, "Rocky Cometary Particulates: Their Elemental Isotopic and Mineralogical Ingredients," *Space Science Reviews* 90: 91-97, 1999.

28. National Research Council, *The Astrophysical Context of Life*, The National Academies Press, Washington, D.C., 2005, pp. 41-42.

29. L. Becker and R.J. Poreda, "Fullerene and Mass Extinctions in the Geologic Record," *Meteoritics and Planetary Science* 36: A17, 2001.

30. K.O. Pope, S.W. Kieffer, and D.E. Ames, "Empirical and Theoretical Comparisons of the Chicxulub and Sudbury Impact Structures," *Meteoritics and Planetary Science* 39: 97-116, 2004.

31. Tholins are organic solids produced by the irradiation of cosmically abundant reducing gases (C. Sagan and B. Khare, "Tholins: Organic Chemistry of Interstellar Grains and Gas," *Nature* 277, 102-107, 1979). "Tholin" is not a specific term, since one will generate a wide array of different solids depending on the type of irradiation used and the composition of the mixture of gases irradiated. It was coined from the Greek word "tholos," meaning "muddy."

32. C.A. Trujillo and M.E. Brown, "A Correlation Between Inclination and Color in the Classical Kuiper Belt," *Astrophysical Journal Letters* 566: 125-128, 2002.

33. O.R. Hainaut, C.E. Delahodde, H. Boehnhardt, E. Dotto, M.A. Barucci, K.J. Meech, J.M. Bauer, R.M. West, and A. Doressoundiram, "Physical Properties of TNO 1996 TO66," *Astronomy and Astrophysics* 356: 1076-1088, 2000.

34. T. Hiroi, M.E. Zolensky, and C.M. Pieters, "The Tagish Lake Meteorite: A Possible Sample from a D-type Asteroid," *Science* 293: 2234-2236, 2001.

35. P.D. Wilson and C. Sagan, "Spectrophotometry and Organic Matter on Iapetus. 1: Composition Models," *Journal of Geophysical Research* 100: 7531-7537, 1995.

36. J. Hunter Waite, Jr., M.R. Combi, W-H. Ip, T.E. Cravens, R.L. McNutt, Jr., W. Kasprzak, R. Yelle, J. Luhman, H. Niemann, D. Gell, B. Magee, G. Flecther, J. Lunine, and W-L. Tseng, "Cassini Ion and Neutral Mass Spectrometer: Enceladus Plume Composition and Structure," *Science* 311: 1419-1422, 2006.

37. J.M. Bauer, T.L. Roush, T.R. Gaballe, K.J. Meech, T.C. Owen, W.D. Vacca, J.T. Rayner, and K.T.C. Jim, "The Near-infrared Spectrum of Miranda: Evidence of Crystalline Water Ice," *Icarus* 158: 178-190, 2002.

38. J. Granahan, "A Compositional Study of Asteroid 243 Ida and Dactyl from Galileo NIMS and SSI Observations," *Journal of Geophysical Research* 107: 20-21, 2002.

39. J.F. Bell, N.I. Izenberg, P.G. Lucey, B.E. Clark, C. Peterson, M.J. Gaffey, J. Joseph, B. Carcich, A. Harch, M.E. Bell, J. Warren, P.D. Martin, L.A. McFadden, D. Wellnitz, S. Murchie, M. Winter, J. Veverka, P. Thomas, M.S. Robinson, M. Malin, and A. Cheng, "Near-infrared Reflectance Spectroscopy of 433 Eros from the NIS instrument on the NEAR Mission. I. Low Phase Angle Observations," *Icarus* 155: 119-144, 2002.

40. J. Veverka, M. Robinson, P. Thomas, S. Murchie, J.F. Bell III, N. Izenberg, C. Chapman, A. Harch, M. Bell, B. Carcich, A. Cheng, B. Clark, D. Domingue, D. Dunham, R. Farquhar, M.J. Gaffey, E. Hawkins, J. Joseph, R. Kirk, H. Li, P. Lucey, M. Malin, P. Martin, L.

McFadden, W.J. Merline, J.K. Miller, W.M. Owen, Jr., C. Peterson, L. Prockter, J. Warren, D. Wellnitz, B.G. Williams, and D.K. Yeomans, "NEAR at Eros: Imaging and Spectral Results," *Science* 289: 2088-2097, 2000.

41. J.I. Trombka, S.W. Squyres, J. Brückner, W.V. Boynton, R.C. Reedy, T.J. McCoy, P. Gorenstein, L.G. Evans, J.R. Arnold, R.D. Starr, L.R. Nittler, M.E. Murphy, I. Mikheeva, R.L. McNutt, Jr., T.P. McClanahan, E. McCartney, J.O. Goldsten, R.E. Gold, S.R. Floyd, P.E. Clark, T.H. Burbine, J.S. Bhangoo, S.H. Bailey, and M. Petaev, "The Elemental Composition of Asteroid 433 Eros: Results from the NEAR-Shoemaker X-ray Spectrometer," *Science* 289: 2101-2105, 2000.

42. R.P. Binzel, A.W. Harris, S.J. Bus, and T.H. Burbine, "Spectral Properties of Near-Earth Objects: Palomar and IRTF Results for 48 Objects Including Spacecraft Targets (9969) Braille and (10302) 1989 ML," *Icarus* 151: 139-149, 2001.

43. B.J. Buratti, D.T. Britt, L.A. Soderblom, M.D. Hicks, D.C. Boice, R.H. Brown, R. Meier, R.M. Nelson, J. Oberst, T.C. Owen, A.S. Rivkin, B.R. Sandel, S.A. Stern, N. Thomas, and R.V. Yelle, "9969 Braille: Deep Space 1 Infrared Spectroscopy, Geometric Albedo, and Classification," *Icarus* 167: 129-135, 2004.

44. D.C. Boice, L.A. Soderblom, D.T. Britt, R.H. Brown, B.R. Sandel, R.V. Yelle, B.J. Buratti, M.D. Hicks, R.M. Nelson, M.D. Rayman, J. Oberst, and N. Thomas, "The Deep Space 1 Encounter with Comet 19P/Borrelly," *Earth, Moon, and Planets* 89: 301-324, 2002.

45. For a more complete description of touch-and-go sampling see, for example, Space Studies Board, National Research Council, *A Scientific Rationale for Mobility in Planetary Environments*, National Academy Press, Washington, D.C., 1999, p. 37.

46. A.S. Rivkin, *Observations of Main-belt Asteroids in the 3-Micron Region*, Ph.D. Thesis, University of Arizona, 1997.

47. A. Coradini, F. Capaccioni, P. Drossart, A. Semery, G. Arnold, U. Schade, F. Angrilli, M.A. Barucci, G. Bellucci, G. Bianchini, J.P. Bibring, A. Blanco, M. Blecka, D. Bockelee-Morvan, R. Bonsignori, M. Bouye, E. Bussoletti, M.T. Capria, R. Carlson, U. Carsenty, P. Cerroni, L. Colangeli, M. Combes, M. Combi, J. Crovisier, M. Dami, M.C. DeSanctis, A.M. DiLellis, E. Dotto, T. Encrenaz, E. Epifani, S. Erard, S. Espinasse, A. Fave, C. Federico, U. Fink, S. Fonti, V. Formisano, Y. Hello, H. Hirsch, G. Huntzinger, R. Knoll, D. Kouach, W.H. Ip, P. Irwin, J. Kachlicki, Y. Langevin, G. Magni, T. McCord, V. Mennella, H. Michaelis, G. Mondello, S. Mottola, G. Neukum, V. Orofino, R. Orosei, P. Palumbo, G. Peter, B. Pforte, G. Piccioni, J.M. Reess, E. Ress, B. Saggin, B. Schmitt, D. Stefanovitch, A. Stern, F. Taylor, D. Tiphene, and G. Tozzi, "Virtis: An Imaging Spectrometer for the Rosetta Mission," *Planetary and Space Science* 46: 1291-1304, 1998.

48. National Research Council, *New Frontiers in the Solar System: An Integrated Exploration Strategy*, The National Academies Press, Washington, D.C., 2003, pp. 31-32, 115, 136, 145, and 206-207.

49. D.P. Cruikshank and C.M. Dalle Ore, "Spectral Models of Kuiper Belt Objects and Centaurs," *Earth, Moon, and Planets* 92: 313-330, 2003.

50. W.M. Grundy and J.A. Stansberry, "Mixing Models, Colors and Thermal Emissions," *Earth, Moon, and Planets* 92: 331-336, 2003.

51. National Research Council, *New Frontiers in the Solar System: An Integrated Exploration Strategy*, The National Academies Press, Washington, D.C., 2003, pp. 206-207.

52. National Research Council, *New Frontiers in the Solar System: An Integrated Exploration Strategy*, The National Academies Press, Washington, D.C., 2003, p. 195.

5

The Giant Planets and Their Satellites

LARGE SATELLITES OF THE OUTER SOLAR SYSTEM

In addition to the gas-giant planets, the outer solar system hosts at least seven objects with radii greater than 1000 km. These objects include Jupiter's four large satellites Io, Europa, Ganymede, and Callisto (often referred to as the Galilean satellites), Saturn's large moon Titan, Neptune's large satellite Triton, and the planet/Kuiper-belt object Pluto. Each of these objects is remarkably diverse and displays a variety of planetary processes, and six (i.e., all except Io) are sites of interest for the study of organic chemistry in the solar system (Table 5.1).

While each object has unique characteristics with regard to organic chemistry in the solar system, they can be arranged in four groups. Jupiter's icy satellites Ganymede, Callisto, and especially Europa each show evidence of layers of liquid water that offer potentially uniquely interesting environment for organic synthesis. These bodies are also subject to intense radiation bombardment at their locations within Jupiter's magnetosphere that can potentially affect carbon chemistry in the surface and near-surface ices. Triton and Pluto, while in very different dynamical circumstances, are near-twins in terms of their bulk properties. Both have nitrogen-dominated atmospheres, significant albedo variations on their surfaces, and comparable surface compositions as inferred from infrared spectra. In a group of its own is the organic-rich satellite Titan. With a dense, cloudy atmosphere that processes methane into complex hydrocarbons which condense and precipitate out on the surface, it may be the richest organic environment in the solar system aside from Earth. Lastly, while Io is a fascinatingly complex object, its vigorous volcanism makes it an inhospitable and unlikely site for organics.

Many of the molecules listed in Table 5.1 have been identified by remote spectroscopic observations at infrared and ultraviolet wavelengths, both from ground- and space-based platforms. In the cases of Triton and Pluto, the existence of atmospheric CO is inferred from the presence of surface ice and the requirement for vapor pressure equilibrium; upper limits are available from direct measurement. CH_4 is thought to be a component of these atmospheres for the same reason and has been directly detected on Pluto.[1] The Galileo Near Infrared Mapping Spectrometer (NIMS) experiment has been particularly important for identifying carbon-containing molecules and functional groups on the surfaces of the icy Galilean satellites.

Sources of Carbon

Europa, Ganymede, and Callisto

The Galilean satellites (Io, Europa, Ganymede, and Callisto) have been studied extensively by both spacecraft flybys and remote telescopic observations. The three so-called icy Galileans—Europa, Ganymede, and Callisto—

72

TABLE 5.1 Observed Carbon Inventory in Large Icy Bodies of the Outer Solar System

Object	Radius (km)	Atmosphere		Surface	Energetics
		Dominant	Carbon		
Europa	1,560	O_2	—	CO_2	Particle radiation, tidal heating, photodissociation?
Ganymede	2,638	O_2	—	C-H, C≡N, CO_2 (organic, –CHO)	
Callisto	2,410	O_2	—		
Pluto	1,137	N_2	CO (<0.06) CH_4 (0.0003-0.0045)	CH_4, CO (red organic)	Photodissociation, solid-state greenhouse?, cosmic rays
Triton	1,352	N_2	CO (<0.59)	CH_4, CO, CO_2 (red organic)	
Titan	2,575	N_2	CH_4 (0.015-0.02) C_2H_2 (10^{-6}) C_2H_6 (10^{-5}) C_3H_8 (10^{-7}) C_2H_4 (10^{-6}-10^{-8}) HCN (10^{-7}) C_4H_2 (10^{-9}) C_3H_4 (10^{-9}) HC_3N (10^{-7}-10^{-9}) C_2N_2 (10^{-9}) CO_2 (10^{-8}) CO (10^{-5})	Organic precipitates, source for atmospheric CH_4 (liquid hydrocarbons?)	Photodissociation, impacts?, cosmic rays

have spectra that are characterized by strong infrared bands of water ice.[2,3] Models indicate that the water ice fraction varies from 20 percent in some dark regions to nearly 100 percent in others. The remaining material, which sometimes makes up the majority of the surface, is a good match with a variety of hydrated silicates similar to the hydrated silicates found in carbonaceous chondrites.[4] In addition, the surfaces have a number of minor constituents that are variable in observability and abundance.

At least three different carbon-bearing molecules have been identified in spectra of the surfaces of these objects. Spectra obtained by Galileo's NIMS show absorption bands indicative of carbon-rich materials. Absorptions are interpreted to indicate C-H and C≡N present in organic compounds. Spectra also indicate CO_2, and its spectral characteristics suggest that it is trapped within or bound to dark surface materials. A broad absorption in the ultraviolet is suggestive of trapped O_3 and another unidentified band that may be the $–CH_2O$ functional group of an organic compound.[5]

Carbon dioxide shows interesting and suggestive correlations with both albedo and topography in the three icy Galileans. Callisto, the outermost of the four satellites, has an ancient, heavily cratered surface with widespread dark material that is thought to be carbon-rich. Galileo images show that Callisto's dark material covers the underlying topography, filling in craters and other topographic lows between bright, ice-rich high-standing knobs and crater rims. CO_2 is correlated with fresh material such as bright impact craters, suggesting that CO_2 is a possible component of Callisto's icy subsurface. CO_2 is strongest on the trailing side of Callisto, suggesting a possible link to irradiation by jovian magnetospheric plasma. Dark heavily cratered terrain constitutes about one-third of Ganymede's surface. Geological investigations using Galileo's high-resolution images suggest that the dark material is a relatively thin layer above brighter icy material and has been affected by processes of sublimation, mass wasting, ejecta blanketing, and tectonism. Dark deposits within topographic lows, such as craters, within grooved terrain have a measured albedo as low as 12 percent. As for Callisto, NIMS observations show absorptions in Ganymede's dark terrain, with bands less intense by about a factor of two than those on Callisto. The CO_2 absorption shows a mottled distribution, generally correlating with the darker terrain. CO_2 has also been detected by Galileo's NIMS on Europa, where it is associated with lower-albedo material.

Titan

In terms of organic chemistry, the premiere destination within the solar system is Saturn's Mercury-sized satellite Titan. Its 1.5-bar atmosphere contains about 98 percent nitrogen and about 2 percent methane and more complex hydrocarbons. In addition, the atmosphere contains small amounts of nitrogen-containing organics as well as a haze that is believed to be composed of hydrocarbons and polymers formed by the action of solar ultraviolet radiation on this atmosphere. The surface plays an integral, if less well observed, role in this complex system, acting as both a sink for condensed organics and a source needed to replenish atmospheric methane as it is processed.

Titan's carbon chemistry begins with methane present in Titan's atmosphere. Photons at λ <160 nm are able to remove a hydrogen atom from CH_4 to produce a methyl radical, $CH_3^{\cdot}$. A wide variety of hydrocarbons are then produced, and many of these have been detected spectroscopically by both Earth- and space-based instruments (see Table 5.1). Magnetospheric electrons from Saturn convert N_2 to N atoms that are then free to react with hydrocarbons to produce nitriles, which are also observed.[6-8] Because of Titan's relatively low mass, H and H_2 produced from the breakup of methane escape to space, enabling the chemical evolution of complex organic compounds.

Titan's orange-colored haze is believed to result from the formation of polymers of C_2H_2, HCN, and C_2H_4 in the atmosphere initiated through additional photochemical and/or charged-particle reactions.[9,10] Many of the observed and predicted organics will condense at the temperatures found in Titan's stratosphere and accumulate on the surface, where they remain sequestered and/or participate in further chemical processing. Indeed, it has been speculated that organic precipitates could be further processed (and additional organic compounds synthesized) on Titan's surface through liquid-phase chemistry simply by condensation and accumulation and/or the influence of impact events.[11,12]

Despite the abundant hydrocarbons observed in the atmosphere and the theoretical predictions of substantial deposits of hydrocarbons on the surface, evidence for liquid or solid organics has been elusive. While Titan's atmosphere is opaque at most wavelengths due to a combination of particle scattering and gas absorption, there are several infrared windows with sufficiently low optical depths which allow probing of the surface at selected wavelengths.[13] These windows have been exploited to detect distinctive, time-stable albedo features on Titan's surface, ruling out a global ocean. Spectrophotometry can also be carried out within these windows, and the results suggest water ice as an important surface component.[14] Earth-based radar observations of Titan are also inconsistent with a global ocean but appear to suggest smaller, relatively smooth features that might be lakes.[15,16]

Significant advances in understanding of the nature of Titan's organics will almost certainly result from ongoing studies by the Cassini-Huygens spacecraft. Cassini is currently carrying out radar and near-infrared spectroscopic and mass-spectral measurements in Titan's atmosphere and observations of the satellite's surface, and the Huygens probe has conducted imaging and in situ chemical data collection.[17,18]

Although Cassini's explorations are far from complete, the successful descent of Huygens through Titan's atmosphere on January 14, 2005, combined with the bonus of an unexpectedly long period of surface observations, has confirmed some long-standing expectations and revealed some intriguing new characteristics of Saturn's largest satellite. Images from Huygens' descent imager, for example, showed features highly reminiscent of drainage channels and shorelines. Similarly, images obtained on the surface show icy pebbles rounded, perhaps, as a result of fluvial activity (see cover illustration). But standing bodies of liquid hydrocarbons were not apparent in any of the images returned by Huygens.

Data from Huygens on the variations of temperature and pressure as a function of altitude were virtually indistinguishable from those expected on the basis of models derived from observations made during Voyager 2's flyby in 1981. However, the atmosphere appears to lack the expected argon (in the form of ^{36}Ar and ^{38}Ar), krypton, and xenon—a possible sign that Titan accreted at a somewhat higher temperature than previously expected. However, the instruments did detect ^{40}Ar, a daughter product of ^{40}K released from Titan's interior, perhaps as a result of cryovolcanic activity. Chemical analyses performed by Huygens' gas chromatograph/mass spectrometer (GC/MS) revealed that methane becomes more abundant relative to nitrogen with the approach closer to Titan's surface. The GC/MS also registered a sharp increase in the methane abundance soon after landing, possibly indicating the presence of liquid methane just below Titan's surface.

The instruments on Cassini itself are also returning important data, the significance of which is still not entirely clear. Images of Titan's surface show few craters, indicating a geologically active world. However, the surface does not show any significant compositional variations. Data from the initial radar investigations of Titan's surface, covering just a few percent of the globe, are intriguing and appear to show structures that appear to be lakes of hydrocarbons.[19] Continued radar mapping, in conjunction with ground truth provided by the Huygens landing, will improve understanding of Titan as an active world.

Another highly unexpected finding is the detection of benzene in Titan's upper atmosphere, as determined by in situ analysis performed during one of Cassini's first close flybys. With so many tantalizing initial findings and with Cassini scheduled to make some dozens of additional Titan flybys during the next several years, it is clear that Titan will be a prime objective for additional studies long after Cassini has ceased operation.

Triton and Pluto

The N_2-dominated atmospheres of Triton and Pluto are generally similar to each other. Trace quantities of CH_4, CO, and Ar are consistent with their respective vapor-pressure equilibria with ices of CH_4, CO, and Ar on the surfaces of these objects. In part because the atmospheres are much less dense than Titan's, however, the range of expected chemical reactions is more limited. The atmosphere of Triton is so thin that both the thermosphere (where ion-molecule reactions are important) and the surface determine its atmospheric composition. Methane in Triton's atmosphere is photolyzed and depleted over short time scales and must be replenished from surface or subsurface reservoirs. The CH_4 photochemistry is believed to be similar to that on Titan, but there is greater uncertainty due to lack of knowledge of the relevant chemical kinetic parameters for the colder temperatures of Triton's atmosphere. Moreover, any complex hydrocarbon and nitrile species will easily condense, forming aerosols near Triton's surface due to temperatures colder than those on Titan.[20]

Triton is Neptune's only large satellite, and its retrograde and highly inclined orbit suggests capture by Neptune. Its surface has relatively few craters, indicating a surface age of only $\sim10^8$ years, which suggests it remains highly active today. The surface shows abundant evidence for tectonism, cryovolcanic activity, and solid-state convection. Active geysers were discovered in Voyager images, with plumes that extend up to ~8 km to the top of the troposphere and are carried by the tenuous winds. The geysers are believed to be powered by N_2 gas heated by geothermal or solar energy, and dark surface streaks, some more than 100 km long, mark the fallout of dark dust carried aloft with the geyser plumes.

Ground-based infrared spectra show that Triton's surface is rich in volatile ices, specifically N_2, CH_4, CO, CO_2, and H_2O. CH_4 and CO are believed to be in solid solution in the more abundant N_2 ice. Additional molecules are predicted as surface precipitates resulting from atmospheric photochemistry, specifically HCN, C_2H_4, C_2H_6, and C_2H_2, but these have not yet been confirmed spectroscopically. Triton has a distinctly pink color, leading to the suggestion that ultraviolet and energetic particle irradiation of CH_4 ice has created organic chromophores. Organic materials are a suspected, but unconfirmed, component of the dark surface streaks associated with Triton's geysers.

Pluto and its moon Charon have not yet been visited by a spacecraft, although they are expected to be by the middle of the next decade by the New Horizons mission. Recent infrared data indicate the presence of N_2, CH_4 (both pure and in an icy matrix), CO, and H_2O. Ground-based mutual satellite occultation data and modeling, as well as more recent Hubble Space Telescope observations, show broad-scale albedo variations on Pluto's surface, providing evidence for dark material in its equatorial regions. As on Triton, organic materials are suspected, produced from energetic processing of surface materials or from atmospheric precipitation. This could explain Pluto's reddish color. Organics are consistent with infrared spectra and could cover approximately 10 to 15 percent of the surface, but these spectra do not require or confirm the presence of organics. Depending on conditions of its formation, Pluto's interior could contain an organic-rich layer up to 100 km thick.

Energetic Processes

Radiolysis and Photolysis in Ices

Organic molecules can be both synthesized and destroyed in the outer solar system by irradiation. (See the section "Synthesis and Destruction of Organic Materials" in Chapter 4.) Galileo data for the jovian satellites suggest the presence of CO_2 and, possibly, CN in the surfaces of the icy satellites, as described above. It is not clear whether that CO_2 is from outgassing or is due to decomposition of carbonates at the surface. However, it is clear that CO_2 in an ice matrix exposed to radiation may lead to simple organics formed on the icy-satellite surfaces.

Radiation processing occuring on such bodies is indicated by the radiation-induced sulfur cycle observed on Europa's surface and the presence of oxygen and peroxide produced from radiolysis of ice.[21,22] The presence of sulfur in H_2O exposed to radiation leads to radiation-induced cycling among three primary forms of sulfur— hydrated H_2SO_4, SO_2, and chain sulfur—at equilibrium concentration ratios of about 10:1:1. Such ratios are roughly consistent with reflectance observations of Europa at infrared, ultraviolet, and visible wavelengths, respectively. Depending on the amount of CO_2 in ice, similar processing can lead to the formation of frozen carbonic acid and, possibly, more complex compounds.

Photolysis in Atmospheres

Ultraviolet light from the Sun and, to a lesser extent, energetic cosmic rays provide non-equilibrium energy sources that can drive pathways of chemical synthesis in the atmospheres of large outer solar system objects. The effects of photochemical atmospheric processes are most apparent on Titan,[23,24] but these processes are also active in the more tenuous atmospheres of Pluto and Triton.[25-27] In all three, CH_4 is the principal source for subsequent organic synthesis. The presence of N_2 allows for the significant production of nitriles, with HCN as the simplest example. Hydrocarbons and nitriles condense in all three of these atmospheres, resulting in hazes. Precipitation then deposits organics on the surfaces, where they can be further processed as detailed in the previous section.

Although the basic outlines of photochemistry in these atmospheres are understood, some species are not adequately modeled. Observations are available to constrain the models for only relatively simple molecules (see Figure 5.1). The uncertainties for larger organic compounds are considerable. Most photochemical models are one-dimensional approximations, and only recently have some models attempted to deal with two-dimensional approximations.[28] A follow-up mission to Titan, for example, should be equipped to study the three-dimensional distribution of a variety of organics through both remote sensing and direct sampling of the atmosphere. Laboratory studies may also be conducted under simulated outer solar system conditions, and photolytic products can be studied directly in the laboratory.[29]

Radiogenic and Tidal Heating

The Galileo orbiter made numerous close flybys of each of Jupiter's large icy satellites, Europa, Ganymede, and Callisto. From tracking of the spacecraft trajectory it is possible to determine the moments of inertia in sufficient detail to constrain the internal structure of these bodies.[30] Both Europa and Ganymede are differentiated whereas Callisto is largely undifferentiated, a fact that points to important differences in their formation and evolution within the jovian satellite system. Magnetic field measurements point to the existence of liquid subsurface oceans in all three satellites beneath ice caps that may be tens to hundreds of kilometers thick.

Radiogenic heating takes place on all of the bodies in the solar system, with varying degrees of impact on the structure of the object. Tidal heating takes place only for objects that participate in orbital resonances. This restriction is probably only structurally important for the three Galilean satellites, Io, Europa, and Ganymede. Both of these sources of heating are relevant to carbon chemistry primarily by their ability to create and maintain liquid

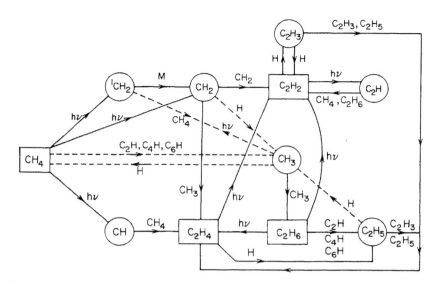

FIGURE 5.1. A schematic representation of the modeled hydrocarbon photosynthetic reaction scheme in Titan's atmosphere. SOURCE: Y.L. Yung, M. Allen, and J.P. Pinto, "Photochemistry of the Atmosphere of Titan: Comparison Between Model and Observations," *Astrophys. J. Suppl.* 55: 465-506, 1984. Copyright 1984. Reprinted with permission of the American Astronomical Society.

environments that are conducive to complex carbon chemistry. Secondarily, environmental niches, perhaps analogous to deep ocean vents on Earth, may provide energy gradients that can directly drive carbon chemistry.[31]

Another important effect of tidal stresses on Europa, in particular, is the creation of cracks in the ice shell that may permit organic enriched material from deeper in the mantle to be transported to the surface where it is accessible for remote observation. Many researchers have noted the asymmetry of the 2.0-μm water band of Europa, an asymmetry that is produced by bound water molecules. While an asymmetry can be a created by hydrated minerals, Dalton[32] argues that details of the band structure are more consistent with biogenic material. On the contrary, Clark[33] argues that the asymmetric water bands are the product of trapped hydronium ions (H_3O^+) that create a hostile environment for organics. While the nature of the non-ice surface material on Europa remains uncertain, the notion of investigating material near surface cracks on Europa remains an important goal for possible future missions, including surface landers of the type that were being considered for the now deferred Jupiter Icy Moons Orbiter (JIMO).[34]

Impacts

Impacts of various sizes may play important roles in organic synthesis on outer solar system bodies. Infrequent but very large impacts may produce transient dense atmospheres and/or lead to large-scale resurfacing with fresh material from the interior of the body. Moderate-sized impactors can melt significant volumes of surface ice that may be sufficient to initiate organic synthesis.[35] Electrical discharges produced by impacts may also result in organic synthesis.[36] Shocks produced by impacts have been shown to result in the formation of polycyclic aromatic hydrocarbons from benzene, a molecule detected during Cassini's studies of Titan and also observed in Jupiter and Saturn.[37]

Micrometeorite impacts can be a significant source of material delivered to both atmospheres and surfaces. For example, micrometeorites are thought to be the main source of oxygen delivered to reduced environments like that at Titan. Micrometeorites may also be a direct source of processed organics, as has frequently been speculated for Earth.

Titan is, perhaps, the most interesting object with respect to the possible role of large impacts. Model calculations predict the accumulation of 100 to 1000 m of solid and liquid hydrocarbons on Titan's surface. If such an accumulation were a permanent sink for methane, Titan's atmospheric methane would be depleted in 10 million years. One proposed solution to the precipitation of organics on the surface and the need for a source of methane to resupply the atmosphere was the theoretical suggestion of a possible global ocean of ethane and methane, which could serve as both a source (methane) and a sink (ethane) for the methane photolytic cycle in the atmosphere. The suggestion of a global ocean appears to be inconsistent, however, with existing observational data.[38-41] Currently, these observations (from ground-based radar, near-infrared, and the Hubble Space Telescope), taken together, suggest that liquid-phase hydrocarbons on Titan's surface are certainly not global but may exist in craters;[42] a supposition now potentially confirmed by Cassini's radar observations.[43] If liquid-phase hydrocarbons on Titan's surface are not sufficient to resupply methane at a steady-state level, it has also been proposed that perhaps Titan is currently in an unusual epoch in its history in which its atmosphere is in an unusually dense state that will eventually become less dense like those found around Triton and Pluto.[44] If such a scenario is true, layered deposits of organics on Titan's surface may preserve a record of these events that would be accessible to advanced missions that might follow Cassini/Huygens.[45]

Solid-State Greenhouse

The Voyager flyby of Neptune and Triton in 1989 revealed the presence of plumes of dark material rising approximately 8 km in the atmosphere, with dark material subsequently blown 100 km downwind leaving surface streaks. The location of the plumes at latitudes receiving maximal seasonal insolation suggested a solid-state greenhouse as the energy source for these geysers.[46,47] Geomorphic evidence[48] favors the geyser model over an alternative "dust-devil" atmospheric transport model.[49] The significance of the relationship between subsurface heating, the dark plume material, and carbon chemistry on Triton is an interesting open question. Although analogs to the Triton plumes have not been found elsewhere in the outer solar system, the New Horizons mission to Pluto will be an important test of the extent of this phenomenon.[50]

Outer Solar System Satellites: Recommendations

Titan Follow-up Mission

Titan is a major reservoir of organic materials. Processes occurring in Titan's atmosphere may provide an ongoing example of the formation of complex abiotic organics from methane, although this example is probably not pertinent to the processes on primitive Earth because, at Titan's 96 K surface temperature, all water is condensed as ice. Due to the complex abiotic organic synthesis, this moon merits continued and close ground-based observation and modeling and laboratory studies of its atmospheric chemistry. The Huygens probe revealed much new information about the composition of Titan's atmosphere as it parachuted to the satellite's icy surface in January 2005 and continued to transmit analyses of the surface organics until radio contact was lost. The 2003 solar system exploration decadal survey singled out a follow-on mission to Titan as a likely priority mission for the decade starting in 2014.[51]

Recommendation: Planning should start now for a follow-up of the Cassini mission to Titan that would include a lander sent to sample the surface, since the complexity of the organics there is expected to be greater than that of organics in its atmosphere. The lander should have the capability of sampling organics that are solids at 96 K as well as those that are liquids. The Titan Explorer mission considered by the solar system exploration decadal survey is a good starting point for this planning.

Galilean Satellites Mission(s)

The likely detection of subsurface oceans on Europa, Callisto, and Ganymede has made these bodies prime targets in NASA's plans to search for biologic processes on solar system bodies. Because the likely environment is covered by kilometers to tens of kilometers of ice, their exploration is problematic. However, the europan surface in particular suggests that there may be an exchange of materials between the surface and subsurface due to geological processes driven by tidal heating. Therefore, plans are underway to search from an orbiting spacecraft for signatures of organic species on the surface or at some depth into the surface that can be probed by an impactor, lasers, or charged particles. The key to success would be the ability, in a relatively intense radiation environment in which destruction of organics is occurring, to study the organic fragments and distinguish delivered organics from intrinsic organics and the occurrence of biotic versus abiotic processes. Such a study would push the envelope on available analytic techniques but would be a critical component of assembling an organic inventory of the solar system.

In the late 1990s and early 2000s, NASA's solar system exploration plans included an Europa Orbiter mission that would undertake flyby observations of Callisto and Ganymede prior to entering orbit about Europa.[52] Although excessive cost growth led to the cancellation of this mission, it did not dampen scientific interest in the study of Jupiter's large, icy satellites. The Europa Geophysical Explorer, a somewhat more elaborate version of the Europa Orbiter, was the highest-priority large mission recommended by the 2003 solar system exploration decadal survey.[53] NASA responded to the survey's recommendation by initiating the development of JIMO, the Jupiter Icy Moons Orbiter mission, the first of a series of advanced-technology spacecraft employing nuclear-electric propulsion systems that would have significantly expanded scientific capabilities compared with previous Europa-mission concepts. JIMO would have conducted global orbital mapping surveys of all three icy satellites, at resolutions of 10 m or better, and might have included a small Europa lander. Organic materials can be studied by making provisions for high-signal-to-noise-ratio spectroscopy at resolutions adequate to discriminate potential carbon-bearing species in both high- and low-albedo regions. JIMO was indefinitely deferred in 2005, and NASA and the planetary science community are currently assessing plans for a more conventional and very much less expensive alternative.[54]

Recommendation: The task group reiterates the solar system exploration decadal survey's findings and conclusions with respect to the exploration of Europa and recommends that NASA and the space science community devise a strategy for the development of a capable Europa orbiter mission and that such a mission be launched as soon as is it is financially and programmatically feasible. Any lander should be equipped with a mass spectrometer capable of identifying simple organics in a background of water and hydrated silicates.

Ground-Based Research

Additional ground-based infrared observations are needed to provide further information concerning the organics on Triton, Pluto, Charon, Centaurs, and Kuiper belt objects. Laboratory spectral studies of the ices of potential hydrocarbon species must be performed to facilitate the detection of organics. These studies should be done on mixtures of candidate organics alone and together with water ice to see if the spectra are perturbed when admixed with other substances. Laboratory studies of the irradiation of these ices should be performed to determine what other organic compounds are formed. These studies should be carried out with pure ices as well as with mixtures of these substances with other organic compounds and with water ice. It is also extremely important that an optical database be developed for compounds of interest. The only reliable identification of species responsible for the features present in the near-infrared spectra of these bodies derives from spectral modeling.

THE GIANT PLANETS

The giant planets, Jupiter, Saturn, Uranus, and Neptune, have atmospheres dominated by molecular hydrogen. CH_4 is the most abundant carbon-containing molecule in the upper atmospheres of these planets at a concentration ranging from about 0.3 percent in Jupiter to 2 percent in Neptune. The abundance of CH_4 relative to H_2 increases with distance from the Sun. In terms of mass, the giant planets are, by far, the largest reservoir of carbon in the solar system, except for the Sun. Jupiter alone contains approximately 3 Earth masses of carbon.

Organic Compounds in the Atmospheres of the Giant Planets

Table 5.2 delineates the current state of knowledge on carbon-containing molecules in the atmospheres of the giant planets. The molecules listed have been identified primarily by remote spectroscopic observations, mainly at infrared and ultraviolet wavelengths, from spacecraft missions (particularly the Voyager 1 and 2, Galileo, and Cassini missions) and space- and ground-based telescopes. In situ measurements were made by the Galileo probe that entered Jupiter's atmosphere in December 1995. Approximate mixing ratios, defined as the number ratio of each carbon compound relative to molecular hydrogen, are indicated in Table 5.2 in parentheses.

The directly observable portions of the atmospheres of the giant planets, their upper troposphere, stratosphere, and mesosphere, are composed mainly of hydrogen and helium with trace quantities of other molecular species. Although the total amount of organic material present in these atmospheres is large, it is diluted at least 1,000-fold by hydrogen and helium. These reducing environments differentiate the organic chemistry that occurs in these atmospheres from the chemistry in others in the present-day solar system. While complex organics are formed in these atmospheres by a variety of processes, they are recycled back to methane on time scales that are short relative to the age of the solar system. Therefore, the organic molecules present in the giant planets provide no history of their solar nebula source, nor do they increase in amounts or complexity with time. This fact can be viewed as both a disadvantage and an advantage. Nevertheless, the relative simplicity of the complex reaction schemes in the atmospheres of the giant planets serves as a test to both understand and model such systems, which may also suggest important conditions for organic synthesis in other reducing environments.

Mechanisms of Formation of Organic Compounds in the Atmospheres of the Giant Planets

Generally, the atmospheric organic chemistry of the giant planets can be characterized as steady-state systems composed of methane and a series of related simple hydrocarbons resulting from the dissociation of methane by one of several sources of energy and the subsequent recombination of these molecular fragments. The process of creating new carbon-carbon bonds in the atmospheres is initiated by the production of the radical $CH_3{}^{\cdot}$ derived from CH_4. Direct photolysis of CH_4 by solar photons with $\lambda < 160$ nm is ubiquitous on the giant planets and drives a complex, non-equilibrium chemistry in their stratospheres.[55-57] In terms of both complexity and observability, this process is the most important source of organic compounds in the giant planets.

The precipitation of trapped ions near the magnetic poles is a second process that drives organic chemical synthesis. It has been observed best on Jupiter, where an extensive polar haze with increased production of hydrocarbons are observed.[58,59] Indeed, estimates of the total organic production indicate that auroral sources and the associated ion-molecule chemistry may dominate the global production of carbon compounds on Jupiter. The magnetic fields of Uranus and Neptune are substantially weaker and have more complex geometries compared with those of Jupiter and Saturn. The contribution of charged-particle precipitation to production of organic molecules on these two planets is not well quantified but is probably less important than the photochemical contribution.

Lightning is a more speculative source of energy that can drive non-equilibrium chemistry. Lightning has been observed on Jupiter in localized latitudinal bands by both the Voyager and the Galileo spacecraft,[60] although

TABLE 5.2 Carbon Compounds Observed in the Atmospheres of the Giant Planets

Planet or Satellite	Main Carbon Compound (mixing ratio)	Trace Carbon Compounds (mixing ratio)
Jupiter	CH_4 (0.003)	C_2H_2 (10^{-7}), C_2H_6 (10^{-6}), C_2H_4 (10^{-9}), C_4H_2 (10^{-10}), C_6H_6 (10^{-9}), CO (10^{-9}), CO_2
Saturn	CH_4 (0.005)	C_2H_2 (10^{-7}), C_2H_6 (10^{-6}), C_2H_4, C_4H_2, C_6H_6, CO (10^{-9}), CO_2
Uranus	CH_4 (0.01)	C_2H_2 (10^{-8}), C_2H_6 (10^{-9}), CO (10^{-8}), CO_2
Neptune	CH_4 (0.02)	C_2H_2 (10^{-8}), C_2H_6 (10^{-6}), CO (10^{-6}), CO_2

the vertical and horizontal distribution of the lightning is unknown. Acetylene is a potential tracer of lightning-induced organic synthesis below the region of the atmosphere where CH_4 can be photodissociated, but there are only limited observations with the vertical resolution needed to establish the contribution of lightning synthesis to the total organic synthesis.

The impact of comet SL9 provided graphic evidence of yet another source of energy that can produce transient impulses of organic synthesis in giant-planet atmospheres. Organic grains totaling some 40 percent of the mass of each impactor were observed at the impact sites.[61] However, spectra of the dark material can be matched with the optical constants of Murchison meteorite material, however, suggesting that at least some of this dust was unaltered material from the impactors rather than synthesized in situ.[62]

Regardless of the energy source—lightning, charged-particle impacts, or photons—the subsequent chemistry of CH_4 results in the production of both saturated (e.g., C_2H_6) and unsaturated (e.g., C_2H_2, C_2H_4) hydrocarbons in these atmospheres (Figure 5.2). Model results suggest that about 70 percent of CH_4 destruction results in the synthesis of higher hydrocarbons, while the remainder regenerates CH_4 through various other photochemical pathways.[63]

Chemical-transport models have been constructed for all four giant planets. These models make testable predictions, including the vertical distribution of photochemical products and the abundances of more complex hydrocarbons that have not yet been observed. The computer models typically rely on a series of reaction rate coefficients, many of which must be estimated or extrapolated from very different laboratory conditions. The lack of appropriate laboratory data for reactions of interest stands as a significant barrier to further progress in modeling.

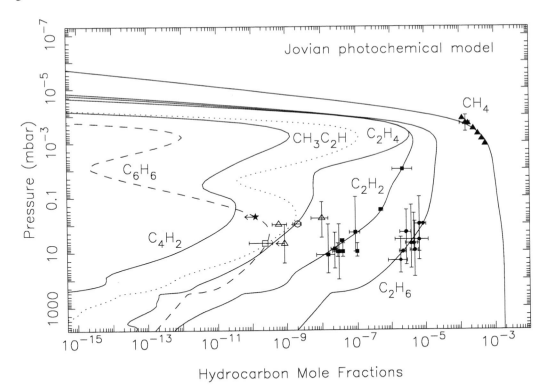

FIGURE 5.2 The mole fractions (P(x)/P(total)) of observed hydrocarbons (symbols) compared to model predictions (curves) as summarized by Moses et al. SOURCE: J.I. Moses, T. Fouchet, R.V. Yelle, A.J. Friedson, G.S. Orton, B. Bezard, P. Drossart, G.R. Gladstone, T. Kostiuk, and T.A. Livengood, "The Stratosphere of Jupiter," in *Jupiter: The Planet, Satellites, and Magnetosphere* (F. Bagenal, T.E. Dowling, and W.B. McKinnon, eds.), Cambridge University Press, 2004. Copyright 2004. Reprinted with the permission of Cambridge University Press.

All four giant planets have stratospheric hazes interpreted as being due to condensed hydrocarbons. These aerosols and cloud particles may also provide sites for further processing of hydrocarbons. The existence of polymers or condensed aromatics has been proposed as a possible source of the unidentified yellow-colored compounds evident in the atmosphere's of Jupiter and Saturn.[64] However, no direct observational evidence to either support or refute this suggestion is available to resolve this long-standing problem. It is possible that an eventual resolution will require in situ sampling of aerosol material with a mass spectrometer.

Below the tropopause, the increasing temperature and pressure with depth drives chemistry away from kinetic control toward equilibrium control. Under equilibrium conditions, ethane is the most abundant hydrocarbon other than methane, present at a mixing ratio of 10^{-9} at a temperature of ~1000 K and pressure of ~800 bar in a solar abundance mixture of gases.[65] Continuing inward to higher temperatures and pressures, CH_4 is replaced by CO as the dominant form of carbon that, in turn, is ultimately replaced by monatomic carbon. It has also been speculated that, at particular depths in giant planet atmospheres, CH_4 may pyrolyze to form latticed carbon. However, it must be noted that much of the modeling of carbon chemistry in the deep atmospheres of giant planets involves poorly characterized reactions in dense, hot, non-ideal fluids of hydrocarbons in a hydrogen-rich background. A better theoretical and experimental understanding of these processes is needed to complete understanding of carbon chemistry in such thick atmospheres.

Overall, the carbon chemistry of the giant planets is essentially a closed system in steady state. The planets are sufficiently massive that loss of hydrogen is not important. There are no abundant solid surfaces on which photochemical products can be stored. Convection eventually transports hydrocarbons synthesized in the stratosphere to deep layers of the troposphere where they are destroyed by pyrolysis and recycled as methane to the upper atmosphere. Thus, no chemical evolution occurs in the system; the chemistry is controlled by kinetics in the upper layers and by thermodynamics in the lower layers.

Advances beyond current understanding of organic synthesis in the atmospheres of giant planets will result from efforts concentrated in three areas:

1. *Laboratory studies of reaction rates.* Modeling of chemical reaction schemes is currently limited by incomplete knowledge of important reaction pathways and rates at temperatures relevant to giant-planet stratospheres.

2. *Remote infrared spectroscopy.* Models of the vertical distribution of photochemical products are far more detailed than available remote sensing measurements capable of constraining these models. Improvements in the state of observations could be obtained by increased opportunities to obtain high-resolution, high-signal-to-noise-ratio spectroscopy (R > 3000, S/N > 100) on large telescopes in the near- and mid-infrared (2-15 μm) capable of high spatial resolution.

3. *In situ sampling.* The unknown identities of the chromophores in the atmospheres of Jupiter and Saturn are a significant gap in knowledge of potential organic chemistry of these planets. Future atmospheric entry probes should consider including experiments designed to identify complex molecules in aerosols and cloud particles that could resolve this long-standing question.

NOTES

1. L.A. Young, J.L. Elliot, A. Tokunaga, C. de Bergh, and T. Owen, "Detection of Gaseous Methane on Pluto," *Icarus* 127: 258, 1997.

2. C.B. Pilcher, S.T. Ridgeway, and T.B. McCord, "Galilean Satellites: Identification of Water Frost," *Science* 178: 1087-1089, 1972.

3. T.B. McCord, G.B. Hansen, and C.A. Hibbitts, "Hydrated Salt Minerals on Ganymede's Surface: Evidence of an Ocean Below," *Science* 292: 1523-1525, 2001.

4. J.M. Moore, C.R. Chapman, E.B. Bierhaus, R. Greeley, F.C. Chuang, J. Klemaszewski, R.N. Clark, J.B. Dalton, C.A. Hibbitts, P.M. Schenk, J.R. Spencer, and R. Wagner, "Callisto," pp. 397-426 in *Jupiter: The Planet, Satellites and Magnetosphere* (F. Bagenal, T.E. Dowling, and W.B. McKinnon, eds.), Cambridge University Press, Cambridge, England, 2004.

5. K.S. Noll, H.A. Weaver, and A.M. Gonnella, "The Albedo Spectrum of Europa from 2200 Å to 3300 Å," *Journal of Geophysical Research* 100(E9): 19057-19060, 1995.

6. D.W. Clarke and J.P. Ferris, "Titan Haze: Structure and Properties of Cyanoacetylene and Cyanoacetylene-acetylene Photopolymers," *Icarus* 127: 158-172, 1997.

7. D.W. Clarke, J.C. Joseph, and J.P. Ferris, "The Design and Use of a Photochemical Flow Reactor: A Laboratory Study of the Atmospheric Chemistry of Cyanoacetylene on Titan," *Icarus* 147: 282-291, 2000.

8. B.N. Tran, J.C. Joseph, J.P. Ferris, P. Persans, and J.J. Chera, "Simulation of Titan Haze Formation Using a Photochemical Flow Reactor: The Optical Properties of the Polymer," *Icarus* 165: 379-390, 2003.

9. D.W. Clarke and J.P. Ferris, "Titan Haze: Structure and Properties of Cyanoacetylene and Cyanoacetylene-acetylene Photopolymers," *Icarus* 127: 158-172, 1997.

10. P. Coll, D. Coscia, N. Smith, M.C. Gazeau, S.I. Ramirez, G. Cernogora, G. Israel, and F. Raulin, "Experimental Laboratory Simulation of Titan's Atmosphere: Aerosols and Gas Phase," *Planetary and Space Science* 47: 1331-1340, 1999.

11. See, for example, D.W. Clarke and J.P. Ferris, "Titan Haze: Structure and Properties of Cyanoacetylene and Cyanoacetylene-acetylene Photopolymers," *Icarus* 127: 158-172, 1997.

12. See, for example, J.I. Lunine, R.D. Lornez, and W.K. Hartmann, "Some Speculations on Titan's Past, Present, and Future," *Planetary and Space Science* 46: 1099-1107, 1998.

13. C.A. Griffith, T. Owen, and R. Wagener, "Titan's Surface and Troposphere, Investigated with Ground-based, Near-infrared Observations," *Icarus* 93: 362-378, 1991.

14. C.A. Griffith, T. Owen, T.R Geballe, J. Rayner, and P. Rannou, "Evidence for the Exposure of Water Ice on Titan's Surface," *Science* 300: 628-630, 2003.

15. D.B. Campbell, G.J. Black, L.M. Carter, and S.J. Ostro, "The Surface of Titan: Arecibo Radar Observations," *EOS Transactions AGU* 84(46), Fall Meeting Supplement, Abstract P42B-08, 2003.

16. D.B. Campbell, G.J. Black, L.M. Carter, and S.J. Ostro, "Radar Evidence for Liquid Surfaces on Titan," *Science* 302: 431, 2003.

17. F. Raulin, P. Coll, D. Coscia, M.C. Gazeau, R. Sternberg, P. Bruston, G. Israel, and D. Gautier, "An Exobiological View of Titan and the Cassini-Huygens Mission," *Advances in Space Research* 22: 353-362, 1998.

18. G. Israel, M. Cabane, P. Coll, D. Coscia, F. Raulin, and H. Niemann, "The Cassini-Huygens ACP Experiment and Exobiological Implications," *Advances in Space Research* 21: 319-331, 1999.

19. E.R. Stofan, C. Elachi, J.I. Lunine, R.D. Lorenz, B. Stiles, K.L. Mitchell, S. Ostro, L. Soderblom, C. Wood, H. Zebker, S. Wall, M. Janssen, R. Kirk, R. Lopes, F. Paganelli, J. Radebaugh, L. Wye, Y. Anderson, M. Allison, R. Boehmer, P. Callahan, P. Encrenaz, E. Flamini, G. Francescetti, Y. Gim, G. Hamilton, S. Hensley, W.T.K. Johnson, K. Kelleher, D. Muhleman, P. Paillou, G. Picardi, F. Posa, L. Roth, R. Seu. S. Shaffer, S. Vetrella, and R. West, "The Lakes of Titan," *Nature* 445: 61-64, 2007.

20. Y.L. Yung and W.B. DeMore, *Photochemistry of Planetary Atmospheres*, Oxford University Press, New York, 1999.

21. R.E. Johnson, R.W. Carlson, J.F. Cooper, C. Paranicas, M.H. Moore, and M.C. Wong, "Radiation Effects on the Surface of the Galilean Satellites," pp. 485-512 in *Jupiter: The Planet, Satellites and Magnetosphere* (F. Bagenal, T. Dowling, and W.B. McKinnon, eds.), Cambridge University Press, Cambridge, England, 2004.

22. R.E. Johnson, T.I. Quickenden, P.D. Cooper, A.J. McKinley, B. Selby, and C. Freeman, "The Production of Oxidants in Europa's Surface," *Astrobiology* 3: 823-850, 2003.

23. L.M. Lara, R.D. Lorenz, and R. Rodrigo, "Liquids and Solids on the Surface of Titan: Results of a New Photochemical Model," *Planetary and Space Science* 42(1): 5-14, 1994.

24. L.M. Lara, E. Lellouch, J.J. López-Moreno, and R. Rodrigo, "Vertical Distribution of Titan's Atmospheric Neutral Constituents," *Journal of Geophysical Research* 101(E10): 23261-23284, 1996.

25. Y.L. Yung, M. Allen, and J.P. Pinto, "Photochemistry of the Atmosphere of Titan—Comparison Between Model and Observations," *Astrophysical Journal Supplement Series* 55: 465-506, 1984.

26. L.M. Lara, W.-H. Ip, and R. Rodrigo, "Photochemical Models of Pluto's Atmosphere," *Icarus* 130: 16-35, 1997.

27. D.F. Strobel, M.E. Summers, F. Herbert, and B. Sandel, "The Photochemistry of Methane in the Atmosphere of Triton," *Geophysical Research Letters* 17: 1729-1732, 1990.

28. S. Leboinnois, D. Toublanc, F. Hourdin, and P. Rannou, "Seasonal Variations of Titan's Atmospheric Composition," *Icarus* 152: 384-406, 2001.

29. See, for example, H. Imanaka, B.N. Khare, E.L.O. Bakes, M.A. Cannady, C.P. McKay, D.P. Cruikshank, J.E. Elsila, R.N. Zare, S. Sugita, and T. Matsui, "Titan's Organic Haze and Condensation Clouds," 35th Meeting of the American Astronomical Society, Division on Planetary Sciences, Abstract 10.04, 2003. Available at http://www.aas.org/publications/baas/v35n4/dps2003/356.htm. Last accessed January 19, 2007.

30. G. Schubert, J.D. Anderson, T. Spohn, and W.B. McKinnon, "Interior Composition, Structure, and Dynamics of the Galilean Satellites," pp. 281-306 in *Jupiter: The Planet, Satellites and Magnetosphere* (F. Bagenal, T.E. Dowling, and W.B. McKinnon, eds.), Cambridge University Press, Cambridge, England, 2004.

31. See, for example, R. Greeley, C.F. Chyba, J.W. Head III, T.B. McCord, W.B. McKinnon, R.T. Pappalardo, and P. Figueredo, "Geology of Europa," pp. 329-362 in *Jupiter: The Planet, Satellites and Magnetosphere* (F. Bagenal, T.E. Dowling, and W.B. McKinnon, eds.), Cambridge University Press, Cambridge, England, 2004.

32. J.B. Dalton, "Infrared Spectra of Extremophile Bacteria Under Europan Conditions and Their Astrobiological Significance," *Bulletin of the American Astronomical Society* 33: 1125, 2001.

33. R.N. Clark, "The Surface Composition of Europa: Mixed Water, Hydronium, and Hydrogen Peroxide Ice," *EOS Transactions AGU* 84(46), Fall Meeting Supplement, Abstract P51B-0445, 2003.

34. See, for example, National Research Council, *Priorities in Space Science Enabled by Nuclear Power and Propulsion*, The National Academies Press, Washington, D.C., 2006, pp. 17-21.

35. H. Imanaka, B.N. Khare, E.L.O. Bakes, M.A. Cannady, C.P. McKay, D.P. Cruikshank, J.E. Elsila, R.N. Zare, S. Sugita, and T. Matsui, "Titan's Organic Haze and Condensation Clouds, 35th Meeting of the American Astronomical Society, Division on Planetary Science, Abstract 10.0405, 2003. Available at http://www.aas.org/publications/baas/v35n4/dps2003/356.htm. Last accessed January 19, 2007.

36. J.G. Borucki, B. Khare, and D. Cruikshank, "A New Energy Source for Organic Synthesis in Europa's Surface Ice," *Journal of Geophysical Research (Planets)* 107: 24, 2002.

37. E.H. Wilson and S.K. Atreya, "Benzene Formation in the Atmosphere of Titan," *Bulletin of the American Astronomical Society* 32: 1025, 2000.

38. D.B. Campbell, G.J. Black, L.M. Carter, and S.J. Ostro, "The Surface of Titan: Arecibo Radar Observations," *EOS Transactions AGU* 84(46), Fall Meeting Supplement, Abstract P42B-08, 2003.

39. C.A. Griffith, "Evidence for Surface Heterogeneity on Titan," *Nature* 364: 511-514, 1993.

40. P.H. Smith, M.T. Lemmon, R.D. Lorenz, L.A. Sromovsky, J.J. Caldwell, and M.D. Allison, "Titan's Surface, Revealed by HST Imaging," *Icarus* 119: 336-349, 1996.

41. R. Meier, B.A. Smith, T.C. Owen, and R.J. Terrile, "The Surface of Titan from NICMOS Observations with the Hubble Space Telescope," *Icarus* 145: 462-473, 2000.

42. D.B. Campbell, G.J. Black, L.M. Carter, and S.J. Oster, "Radar Evidence for Liquid Surfaces on Titan," *Science* 302: 431, 2003.

43. E.R. Stofan, C. Elachi, J.I. Lunine, R.D. Lorenz, B. Stiles, K.L. Mitchell, S. Ostro, L. Soderblom, C. Wood, H. Zebker, S. Wall, M. Janssen, R. Kirk, R. Lopes, F. Paganelli, J. Radebaugh, L. Wye, Y. Anderson, M. Allison, R. Boehmer, P. Callahan, P. Encrenaz, E. Flamini, G. Francescetti, Y. Gim, G. Hamilton, S. Hensley, W.T.K. Johnson, K. Kelleher, D. Muhleman, P. Paillou, G. Picardi, F. Posa, L. Roth, R. Seu. S. Shaffer, S. Vetrella, and R. West, "The Lakes of Titan," *Nature* 445: 61-64, 2007.

44. R.D. Lorenz, C.R. McKay, and J.I. Lunine, "Photochemically Driven Collapse of Titan's Atmosphere," *Science* 275: 642-644, 1997.

45. J.I. Lunine, R.D. Lornez, and W.K. Hartmann, "Some Speculations on Titan's Past, Present, and Future," *Planetary and Space Science* 46: 1099-1107, 1998.

46. R.H. Brown, T.V. Johnson, R.L. Kirk, and L.A. Soderblom, "Energy Sources for Triton's Geyser-like Plumes," *Science* 250: 431-435, 1990.

47. R.L. Kirk, L.A. Soderblom, and R.H. Brown, "Subsurface Energy Storage and Transport for Solar-powered Geysers on Triton," *Science* 250: 424-429, 1990.

48. S.M. Metzger, "Geomorphic Tests of the Geyser and Dust Devil Models for Triton's Plumes," *Lunar and Planetary Science* 27: 871, 1996.

49. A.P. Ingersoll and K.A. Tryka, "Triton's Plumes—The Dust Devil Hypothesis," *Science* 250: 435-437, 1990.

50. P.E. Meade, *Surface-Atmosphere Interaction Processes on the Icy Satellites Io and Triton*, Ph.D. dissertation, University of Colorado, Boulder, 1995.

51. National Research Council, *New Frontiers in the Solar System: An Integrated Exploration Strategy*, The National Academies Press, Washington, D.C., 2003, pp. 132-133 and 197.

52. National Research Council, *A Science Strategy for the Exploration of Europa*, National Academy Press, Washington, D.C., 1999, pp. 11-12.

53. National Research Council, *New Frontiers in the Solar System: An Integrated Exploration Strategy*, The National Academies Press, Washington, D.C., 2003, p. 4.

54. National Research Council, *Priorities in Space Science Enabled by Nuclear Power and Propulsion*, The National Academies Press, Washington, D.C., 2006, pp. 17-20.

55. J.I. Moses, E. Lellouch, B. Bézard, G.R. Gladstone, H. Feuchtgruber, and M. Allen, "Photochemistry of Saturn's Atmosphere," *Icarus* 145: 166-202, 2000.

56. M.E. Summers and D.F. Strobel, "Photochemistry of the Atmosphere of Uranus," *Astrophysical Journal* 346: 495-508, 1989.

57. P.N. Romani, J. Bishop, B. Bézard, and S. Atreya, "Methane Photochemistry on Neptune: Ethane and Acetylene Mixing Ratios and Haze Production," *Icarus* 106: 442-463, 1993.

58. S.J. Kim, J. Caldwell, A.R. Rivolo, R. Wagener, and G.S. Orton, "Infrared Polar Brightening on Jupiter: III—Spectrometry from the Voyager 1 IRIS Experiment," *Icarus* 64: 233-248, 1985.

59. J.A. Friedson, A. Wong, and Y.L. Yung, "Models for Polar Haze Formation in Jupiter's Stratosphere," *Icarus* 158: 389-400, 2002.

60. S.J. Desch, W.J. Borucki, C.T. Russell, and A. Bar-Nun, "Progress in Planetary Lightning," *Reports on Progress in Physics* 65: 955-997, 2002.

61. J. Harrington, I. dePater, S.H. Brecht, D. Deming, V. Meadows, K. Zahnle, and P.D. Nicholson, "Lessons from Shoemaker-Levy 9 About Jupiter and Planetary Impacts," pp. 159-184 in *Jupiter: The Planet, Satellites and Magnetosphere* (F. Bagenal, T.E. Dowling, and W.B. McKinnon, eds.), Cambridge University Press, Cambridge, England, 2004.

62. P.D. Wilson and C. Sagan, "Nature and Source of Organic Matter in the Shoemaker-Levy 9 Jovian Impact Blemishes," *Icarus* 129: 207-216, 1997.

63. J.I. Moses, T. Fouchet, R.V. Yelle, A.J. Friedson, G.S. Orton, B. Bezard, P. Drossart, G.R. Gladstone, T. Kostiuk, and T.A. Livengood, "The Stratosphere of Jupiter," pp. 129-158 in *Jupiter: The Planet, Satellites and Magnetosphere* (F. Bagenal, T.E. Dowling, and W.B. McKinnon, eds.), Cambridge University Press, Cambridge, England, 2004.

64. F.W. Taylor, S.K. Atreya, Th. Encrenaz, D.M. Hunten, P.G.J. Irwin, and T.C. Owen, "The Composition of the Atmosphere of Jupiter," pp. 59-78 in *Jupiter: The Planet, Satellites and Magnetosphere* (F. Bagenal, T.E. Dowling, and W.B. McKinnon, eds.), Cambridge University Press, Cambridge, England, 2004.

65. K. Lodders and B. Fegley, "Atmospheric Chemistry in Giant Planets, Brown Dwarfs, and Low-mass Dwarf Stars: I. Carbon, Nitrogen, and Oxygen," *Icarus* 155: 393-424, 2002.

6

The Terrestrial Planets

The terrestrial bodies of the inner solar system include Mercury, Venus, Earth, Earth's Moon, and Mars. These bodies are composed primarily of rocky material, relatively devoid of carbon and other volatile elements compared to the outer planets of the Sun (Table 6.1). Despite the relative paucity of carbon on Earth, however, it is clear from geological evidence that rich organic environments existed early in Earth's history and, by inference, perhaps on other inner solar system bodies as well.

TABLE 6.1 Comparison of Cosmic Composition and Earth's Crust for an Abridged List of the Lighter Elements

Element	Relative Cosmic Abundance	Percentage of non-H Fraction	Percentage of Earth's Crust	Percent Depletion Gas Phase: Diffuse Clouds
Hydrogen	300,000	—	0.22	0
Carbon	100	24.7	0.19	37
Nitrogen	30.9	7.6	0.002	0
Oxygen	235	58.2	46.6	28
Fluorine	0.01	0.0025	—	—
Sodium	0.6	0.15	2.8	—
Magnesium	10.6	2.6	2.1	76
Aluminum	0.8	0.19	8.1	—
Silicon	9.9	2.4	27.1	91
Phosphorus	0.1	0.02	—	—
Sulfur	5.1	1.3	—	0
Chlorine	0.12	0.02	—	—
Argon	2	0.5	—	—
Potassium	0.03	0.007	2.6	—
Calcium	0.8	0.19	3.6	100[a]
Titanium	0.04	0.01	0.46	100[a]
Chromium	0.18	0.04	—	99
Iron	8.9	2.2	5.0	100[a]
Nickel	0.5	0.12	—	100[a]

[a]Less than 1 percent of cosmic abundance observed in gas phase of diffuse clouds.

The record of life in Earth's early crust,[1] the isotopic geochemical history,[2] and inferences drawn from the lunar impact record[3] all combine to constrain the time frame for the earliest emergence of organic environments and life to sometime between 3.5 billion and 3.9 billion years ago. Recently, evidence of isotopically light carbon, which may be indicative of biologically mediated processes, was measured in some highly metamorphosed rocks from the Isua and Akilia formations (West Greenland), suggesting that organic environments and life may already have existed 3.8 billion years ago.[4,5] However, that evidence is compromised because thermal processes can also cause stable isotope fractionation, and those rocks have been deeply buried and heated at least once, and more likely, many times. If organic matter and life were indeed present some 3.8 billion years ago, then this would place the origins of life within the final stages of the late heavy bombardment of the inner solar system,[6] thus narrowing the window of time needed for life to begin and providing a means both to destroy organic environments and to deliver extraterrestrial organic material to the surfaces of the inner planets. In situ synthesis of organic compounds on the terrestrial planets versus exogenous delivery of extraterrestrial organic material is discussed below in assessing the inventories of organic compounds and in a discussion of mechanisms of formation of organic compounds.

INVENTORY OF ORGANIC COMPOUNDS ON THE TERRESTRIAL PLANETS

Atmospheres

As with the atmospheres of the outer solar system bodies, the organic molecules in the atmospheres of the terrestrial planets, apart from Earth, listed in Table 6.2 have been identified primarily by remote spectroscopic observations, mainly at infrared and ultraviolet wavelengths, from spacecraft missions and space- and ground-based telescopes. In situ and sounding measurements have been obtained for Venus (Mariner, Pioneer Venus, Venera), Mars (Mariner, Viking, martian meteorites), and, of course, Earth. Approximate mixing ratios for the carbon compounds are indicated in Table 6.2 in parentheses.

Surfaces

The surfaces of the inner solar system bodies provide a wide range of conditions, both environmental and geological, where organic compounds may be present. The following sections assess the likelihood of finding organics on the surfaces of the terrestrial planets.

Potential inventories of organic materials on Earth's Moon are of considerable scientific interest. At first sight, the Moon seems an unlikely location for organics. The Moon formed from the crystallization of high-temperature silica melts. Any organic carbon that may have been contained within the precursor material would be converted to simpler organics and H_2 at temperatures as high as 1500 K. Indeed, samples returned from the Moon by the Apollo astronauts and the former Soviet Union's robotic Luna missions are devoid of organic materials above and beyond that expected from the infall of carbonaceous meteorites and traces carbon implanted by the solar wind. Inorganic carbon is also found, for example, at concentrations of some 200 parts per million in lunar fines. Most of this carbon is in the form of carbon monoxide bubbles trapped in lunar glasses, consistent with the idea that the carbon was oxidized by mineral oxides at a high temperature.[7]

TABLE 6.2 Carbon Compounds Observed in Inner Solar System Atmospheres

Class	Planet	Main Carbon Compound (mixing ratio)	Trace Carbon Compound (mixing ratio)
N_2-dominated atmospheres	Earth	CO_2 (0.00037)	CH_4 (10^{-6}) CO (10^{-7} to 10^{-8})
CO_2-dominated atmospheres	Venus	CO_2 (0.96)	CO (10^{-6}) COS (10^{-7})
	Mars	CO_2 (0.95)	CO (10^{-4})

The Moon is of interest to the study of organic environments for two very different reasons:

- The lunar surface as a witness plate. That is, it is a location that provides for long-term integration of collected material and thus might have sampled other carbonaceous asteroids that are not present in recent meteorite collections; and
- The lunar surface as the abode of special microenvironments. The lunar materials studied to date come from the Moon's equatorial regions, and these areas are not typical of all lunar environments.

The second possibility is of considerable potential interest, and the remainder of this section is devoted to its discussion.

Permanently shadowed regions exist at both lunar poles. As long ago as 1961, Watson, Murray, and Brown suggested that the extremely low temperatures experienced in these locations, less than some 50 K, would act as cold traps for volatile material impacting the lunar surface.[8] Thus, for example, water and other volatile materials—derived from comets, asteroids, meteorites, or interplanetary dust particles impacting the Moon's surface or, alternatively, created during the reduction of lunar regolith by H⁻ ions from the solar wind—could freeze out on grains in the polar regions and, in principle, persist for considerable periods of time.[9] Such informed speculation has been supported by the subsequent detection of hydrogen concentrations in the lunar polar regions with the neutron spectrometer on the Lunar Prospector spacecraft.[10] That is readily, but not definitively, explained as ice deposits.

The possibility of water ice deposits at the lunar poles raises the issue of the presence of other volatiles, including organic volatiles, since the likely sources of the water, particularly from comets, may also be abundant sources of organic materials. Given a source of raw materials and the availability of likely energy sources (e.g., from cosmic rays and interstellar ultraviolet radiation), it is reasonable to ask if organic synthesis is actively occurring at the lunar poles.

The irradiation of carbon-, hydrogen- and oxygen-bearing ices by ultraviolet radiation or cosmic rays can lead to the synthesis of organic compounds. Similarly, organics may be formed at the lunar poles by the action of the solar wind on the ice there in the same way that they are formed in ice on interstellar dust particles. The radicals formed by the radiation may react with the inorganic carbonaceous condensates to generate simple organic compounds (see in Chapter 2, in the section "The Interstellar Medium," the subsection "The Synthesis of Interstellar Molecules").[11]

Instruments on NASA's forthcoming Lunar Reconnaissance Orbiter (LRO), scheduled for launch in 2008, will directly address questions relating to polar ices. These instruments include the following:

- The Lunar Exploration Neutron Detector (LEND), which will map the flux of neutrons from the lunar surface to create 5-km-resolution maps of the hydrogen distribution and characterize the surface distribution and column density of near-surface water ice deposits;
- The Diviner Lunar Radiometer Experiment, which will map the temperature of the entire lunar surface at 300-m horizontal scales to identify cold traps and potential near-surface and exposed ice deposits; and
- The Lyman-Alpha Mapping Project (LAMP), which will observe the entire lunar surface in the far ultraviolet to search for exposed surface ices and frosts in the polar regions and will provide subkilometer-resolution images of permanently shadowed regions at the lunar poles.

None of these instruments, nor those on other planned lunar orbiters, such as India's Chandrayaan 1, China's Chang'e, or Japan's Selene or Lunar-A, will directly address key questions surrounding the putative existence of organic materials at the lunar poles. Indeed, it is not clear that the definitive detection and study of lunar organics are possible within the current generation of remote-sensing instruments. It is possible that the secondary payload on LRO's launch vehicle, the Lunar Crater Observation and Sensing Satellite (LCROSS), may return spectroscopic evidence of the presence of organic materials in the Moon's polar regions, but it is likely that the study of lunar organics is more appropriately addressed by a lander mission.

The NRC's solar system exploration decadal survey gave very high priority to a mission designed to collect and return samples from the Moon's South Pole-Aitken basin (SPA). This mission is, however, designed to address questions relating to the absolute chronology of the lunar surface, the timing of the late heavy bombardment and the impact frustration of the origin(s) of life on Earth, and the history of lunar differentiation.[12] The search for and study of lunar organics, though intriguing, does not have the same scientific priority as the goals addressed by the proposed SPA mission. The study of volatiles and organics is, nevertheless, a major scientific theme identified in the solar system exploration decadal survey.[13] Therefore, the inclusion of an in situ organic detection instrument should be regarded as an important adjunct to, but not necessarily a driver of, a future polar lander mission.

Mercury

The illuminated surface of Mercury is too hot (>700 K) to preserve complex organic carbon compounds. Possibly early in its history when the luminosity of the young Sun was approximately 30 percent less than its present value, there may have been a period when exogenous organic carbon delivered during the late heavy bombardment phase of planetary accretion might have accumulated on the surface. However, as the surface temperature rose to the current high values, this carbon would have been pyrolyzed, yielding various volatile molecules (e.g., CO_2, CH_4, C_2H_6) and a highly aromatic (if not graphitic) char. It appears likely that volatile products of such pyrolytic reactions would be lost to the minimal mercurian atmosphere. Pristine exogenous organic carbon might, however, survive in one environment, the bottoms of deep craters at the poles. Given that Mercury has essentially no atmosphere, any surface environment that is not directly illuminated (e.g., deep craters at the poles) experiences extremely cold ambient temperatures (e.g., the temperatures at night on Mercury may drop as low as 100 K). Therefore, it is likely, that complex organic carbon could persist and accumulate in the regolith at such locations. Note that a similar argument can be made for the Moon. Furthermore, the ongoing MESSENGER mission to Mercury does have a gamma-ray spectrometer to observe ice at the poles.

Venus

By virtue of its distance from the Sun, Venus had a larger inventory of organic carbon than Mercury. In fact, it was proposed that extensive radial mixing of protoplanetary material across a wide region of the accretionary disk may have resulted in an initial volatile content and composition of Venus similar to that of Mars and Earth.[14] Whether that proposal is correct or not, subsequent processes of planetary evolution have given rise to Venus's hot and corrosive atmosphere. Currently, the atmospheric pressure is approximately 90 bar. The atmosphere is dominated by CO_2, where extensive green-house warming leads to temperatures on the order of 700 K near the surface. Given this extreme thermal boundary condition, the accumulation and preservation of significant organic carbon is unlikely. It is also unlikely that any exogenous organic carbon would be preserved on the surface due to these extreme conditions.

In order for any endogenous organic matter to be detectable today it should be sequestered in the subsurface. Unfortunately, the extremely high surface temperatures create a thermal boundary condition that precludes the existence of cooler subsurface regions in which thermally labile organic carbon might be found. More thermally stable compounds such as methane and benzene might survive in subsurface reservoirs and/or fluid inclusions.

Earth

The current inventory of organic matter on Earth is dominated by biological sources, in particular the structural biopolymers of vascular plants, i.e., cellulose and lignin. While the total mass of the active biological component is estimated to be $\sim 10^{13}$ kg,[15] the majority of organic carbon lies preserved within sedimentary rocks. Recent estimates place the sedimentary carbon at $\sim 10^{19}$ kg, distributed predominantly within oil shales and coal-bearing strata.[16-18] The source of such organic matter originates from the selective preservation of biomolecular compounds derived predominantly from microbiota and vascular plants.

Reentry of sedimentary organic carbon back into the global carbon cycle occurs in essentially two different ways. First, uplift of organic carbon-rich sedimentary rocks by tectonic processes may lead to surface exposure and erosion. The previously sequestered organic carbon is thus susceptible to oxidative, photochemical, or microbial degradation. Alternatively, the entire sedimentary section may be buried deeper, resulting in progressive thermal metamorphism, whereby the organic carbon entrained in the sediments is thermally converted initially to petroleum and ultimately to methane and various forms of inorganic carbon (e.g., CO_2 and graphite). Sediment subducted at active ocean/continent margins (e.g., the west coast of the Americas) provides a means by which this residual carbon can reenter the atmosphere via volcanic exhalations. Note that at high temperatures, CO_2, CO, and H_2 are stable relative to methane. Subduction and volcanism associated with melting of igneous and sedimentary rocks provides a conduit through which organic carbon is recycled back into the atmosphere.

Abiotic sources of organic carbon currently include the persistent rain of exogeneous organic carbon derived from carbonaceous chondritic meteorites, interplanetary dust particles, and the occasional comet. It has been estimated that early in Earth's history (~4.5 billion years ago) up to ~10^9 kg/yr of organic carbon was delivered to Earth.[19] Although this amount has tailed off considerably, current estimates for the influx of carbon-containing exogenous material is on the order of 2×10^8 kg/yr. Earth may also have received a portion of its volatiles from comets, potentially providing abiotic organic matter which some authors have argued is relevant to the origins and/or evolution of life.[20-22] Impacts may also have shock-synthesized organics in the atmosphere or as a result of the impact event (i.e., impact plume syntheses). Note that impacts would also destroy or modify organic matter. As has been observed in carbonaceous chondrites, the concentration of simple organic molecules under aqueous conditions would ultimately result in the formation of some of the more complex organic compounds (e.g., amino acids, nucleic acid bases, and sugars) typically found in modern cells.[23] However, further studies of cosmogeochemical samples, coupled with laboratory experiments, are needed to probe the degree of chemical complexity that can be attained as a result of exogenous delivery of both intact and perhaps synthesized organic compounds to early Earth, early Mars, and other habitable zones.

Endogenous abiotic production of simple organic compounds currently occurs in volcanic fumaroles (e.g., methane) and/or deep-sea hydrothermal vents (e.g., methane and formic acid).[24,25] Traces of abiotically derived hydrocarbons have been identified in hard rock mines associated with ancient volcanogenic massive sulfide deposits, e.g., the Kidd Creek mine.[26] Laboratory experiments have also demonstrated abiotic synthesis in aqueous media of more complex organic compounds from CO, e.g., long-chain fatty acids, fatty alcohols, and unsaturated hydrocarbons (i.e., lipids for primitive membranes),[27] as well as di- and tricarboxylic acids.[28] Furthermore, condensation reactions of fatty acids with alcohols and amines to form esters and amides also proceed in aqueous high-temperature fluids.[29] This is important for prebiotic micelle formation. It is difficult to assess precisely how much abiogenic organic carbon could accumulate and contribute to Earth's total inventory of organic carbon. One estimate proposes that, based on purely thermodynamic grounds, hydrothermal vent systems could provide up to 10^8 to 10^9 kg of organic carbon per year.[30] Actual measurements of the concentration of methane in hydrothermal plumes emitted along the mid-Atlantic ridge yield methane at concentrations of up to 50 nmol/kg of fluid collected.[31]

Mars

Mars could be the most interesting of the terrestrial planets besides Earth in terms of its potential inventory of organic carbon. By virtue of its distance from the Sun, Mars is expected to have formed from volatile-rich materials and also received volatile-rich exogenous complex organic matter after planetary accretion. Moreover, the current conditions of low temperature, no liquid surface water, low partial pressures of oxygen, and an apparently dormant tectonic state would be expected to provide a good environment for the preservation and accumulation of complex organic carbon absent the ubiquitous oxidizing materials found in the upper-most layers of the martian regolith. Mars has a lower atmospheric entry velocity for infalling debris because of its surface gravity is lower than that of Earth. The mass of meteoritic debris that survives martian atmospheric entry without melting has been estimated to be 8.6×10^6 kg/yr.[32] Because Mars is less tectonically active than Earth, its surface may have accumulated debris over longer periods of time. Organic compounds contained within exogenous debris could have been protected from oxidation, especially once this material was incorporated into rocks and sediments. Although a small amount of organic matter is found in martian meteorites, the Viking lander experiments found no organic matter in the martian regolith. Similarly, the Alpha-Proton-X ray Spectrometer (APXS) experiment on Mars Pathfinder was unable to detect carbon in any form in the martian regolith. It has been proposed that oxidants in the martian regolith oxidize any exogenous or endogenous carbon contained within the near surface.[33] If so, Mars may retain organic carbon deeper within its subsurface, i.e., below the level at which eolian and other "gardening" processes disturb the regolith.

In the absence of an active biosphere, the most important endogenous source of organic matter on Mars is probably the abiotic production of organic compounds (e.g., hydrocarbons) derived from the catalytic reduction of magmatic CO_2. Organosynthesis could have occurred both during volcanic exhalations and by hydrothermal alteration of basaltic crust early in martian history. Interest in all of these possibilities has greatly increased by recent claims of the spectroscopic detection of methane in the planet's atmosphere by both ground-based telescopes,[34,35] and spacecraft observations.[36]

MECHANISMS FOR FORMATION OF ORGANIC COMPOUNDS ON THE TERRESTRIAL PLANETS

The Atmosphere of Prebiotic and Pre-photosynthetic Earth

The composition of the earliest atmosphere on Earth and its evolution have not been precisely reconstructed. Competing theories exist for predictions of the ratio of CO_2 to CH_4 and the presence or significance of NH_3. Actual levels probably depended on details of planetary accretion, interactions between the crust and mantle with outgassed volatiles, the ultraviolet flux from the Sun (which could rapidly destroy CH_4 and NH_3), and rates of outgassing of chemical compounds into the atmosphere.[37,38] Obviously, the atmospheric composition and the ultraviolet flux will determine the degree to which organic molecules could have been synthesized in situ in Earth's early atmosphere.[39]

Early experiments to test whether there could have been a source of atmospherically derived organic carbon delivered to Earth's surface were performed by Urey and Miller, who used an electric discharge to initiate chemical reactions in a gaseous mixture of CH_4, H_2O, H_2, and NH_3.[40,41] These experiments showed that amino acids and other organic acids could be readily produced abiotically in such an atmosphere. If any of the terrestrial planets had such a reduced atmosphere, it appears likely that synthesis of organic compounds might have been a significant factor in generating a surface inventory of organic compounds. However, such a reduced composition for Earth's early atmosphere is now considered unlikely due to the rapid photolysis of CH_4 and NH_3 in an atmosphere without an ultraviolet shield. Moreover, volcanic outgassing would most likely give the early Earth an atmosphere consisting of CO_2 and N_2 rather than CH_4 and NH_3. This, together with independent geochemical and cosmochemical constraints on CO_2 and CH_4 abundances from 2 billion to 4 billion years ago, suggests that the likely composition of Earth's early atmosphere was predominantly N_2 and CO_2.[42,43] Subsequent experiments performed on CO_2/N_2 atmospheres (with and without small amounts of CH_4 and NH_3) have shown that the yield of organic compounds via spark discharge is considerably less than in a highly reduced atmosphere containing mostly CH_4 and NH_3.[44]

Other possibilities for the generation of organic compounds via atmospheric chemistry have also been explored. An intriguing mechanism involves the synthesis of HCN via photochemical reactions between CH_4 (of volcanic origin) and N_2.[45] If early in Earth's history the mantle was much more reduced than it is currently, then the amount of CH_4 emitted from volcanoes would have been greater. Rainout of substantial quantities of HCN would make the subsequent synthesis of purines, pyrimidines, and amino acids possible in the aqueous phase.[46] Similarly, photolytic reactions initiated from CO_2 and H_2O at ultraviolet wavelengths have been postulated to produce rainout of formaldehyde (CH_2O)[47] with the possibility of subsequent condensation reactions yielding primitive sugars. This assumes that the concentration of formaldehyde rose high enough in aqueous solution on Earth's surface.[48]

Earth's Current Atmosphere

Earth's atmosphere is unique in the solar system due to the large amount of O_2 produced biotically by photosynthesis. Unlike Venus, most of Earth's carbon inventory is sequestered in carbonate rocks. The abundance of O_2 in Earth's atmosphere (21 percent) results in the photochemical production of a very thin but important layer of ozone in the stratosphere. This layer absorbs ultraviolet photons in a region of the solar spectrum where there are no other absorbers (and where substantial damage to organic molecules can occur upon absorption).

The Earth's atmosphere is highly oxidizing due to the presence of both O_2 and water vapor (which leads to the formation of highly reactive OH radicals). However, despite the oxidizing capacity of the atmosphere, significant amounts of CH_4 (present at 1.7 parts per million by volume) are present in the atmosphere, principally as a result of metamorphic and biological processes. Given that the chemical lifetime of this CH_4 in Earth's oxidizing atmosphere is short (i.e., days to years), and the atmospheric composition precludes its synthesis by photochemistry, the continual, albeit trace, presence indicates a steady-state CH_4 flux into the atmosphere. This input is predominantly from the biosphere (e.g., methanogens produce much of the CH_4 present in the atmosphere). Thus, while CO and hydrocarbons in the outer planets' atmospheres are abiotic in origin, the CO and CH_4 in Earth's atmosphere result from biological processes and/or human activities such as combustion of fossil biogenic material. Again, these inputs are not stable in an O_2 atmosphere and require constant fluxes to make up for their rapid photochemical/oxidative degradation.

In today's atmosphere, the dominant chemical composition of 21 percent O_2 and 78 percent N_2 also prevents lightning-induced or ion-molecule reactions from producing organic compounds. Lightning produces significant amounts of nitrogen oxides, but no species with carbon-carbon bonds.

The CO_2-Dominated Atmospheres of Venus and Mars

The atmospheres of Venus and Mars are dominated by CO_2. Both are composed of approximately 95 percent CO_2, with most of the remainder N_2. Under those conditions, CO_2 is easily photolyzed to form CO and O. Due to the presence of trace gases, particularly water vapor, the reformation of CO_2 is catalyzed by the reactions of CO and O with radicals such as OH and HO_2 (e.g., $CO + OH \rightarrow H + CO_2$), making CO_2 highly stable in their atmospheres. As a result, more complex carbon-bearing species are not produced in the atmospheres of either Mars or Venus.[49]

Ion-molecule reactions and electrical discharges also do not initiate any further carbon chemistry in either planet's atmosphere. Thus, the only carbon-bearing species observed in the martian atmosphere are CO_2 and CO. On Venus, COS has been observed in addition to CO_2 and CO, and is thought to be produced at the surface by equilibrium reactions between CO_2, CO, and FeS_2 at the high temperature and pressure there. (In contrast, COS is produced predominantly biotically on Earth by marine organisms, although COS is also detected in volcanic gaseous emissions.)

Despite the lack of in situ production of organics in the atmospheres of Mars and Venus, some authors have suggested that chemical disequilibrium between trace constituents of Venus's atmosphere is evidence for microbial life in the planet's lower cloud layers.[50,51] In particular, supporters of this conjecture point to the coexistence of chemical species not normally associated, such as H_2 and O_2 and H_2S and SO_2, and the existence of relatively benign atmospheric regions (i.e., with temperatures between 300 and 350 K, pressures of 1 bar, and water vapor concentrations of several hundred ppm).[52] Such organisms presumably evolved when Venus's climate was more Earth-like and then migrated to the clouds as the planet lost its surface water. Irrespective of such speculations, the evolution and present states of the atmospheres of Venus and Mars still bear on the history and evolution of both biotic and abiotic organic compounds in the solar system. For example, given the similar location in the solar nebula of Mars, Earth, and Venus, they should all have had similar bulk chemical compositions 4.5 billion years ago and would have been exposed to similar early radiation processes. The extent to which their atmospheres have evolved and diverged since that time yields information on the evolution of Earth's atmosphere and the couplings of atmospheric composition with biology/life. Mars and Venus may also provide clues to the composition of past atmospheres on Earth that ultimately would have influenced the distribution of organic compounds on Earth (i.e., carbon reservoirs in the atmosphere compared with those at the surface, in the interior, in the ocean, and so on).

The present atmospheric compositions of Venus and Mars also provide guidance as to the likelihood of finding organics produced abiotically or by ancient biota: Mars, for example, has no effective ultraviolet shield, whereas Earth and Venus do, the former via its significant ozone layer and the latter due to absorbance by sulfur compounds in the upper atmosphere. Therefore, ultraviolet-labile organics deposited on the surface of Mars as a result of either biotic or abiotic processes in the past will have been destroyed by ultraviolet light. Moreover, the photolysis of H_2O by ultraviolet light reaching the martian surface would generate OH and HO_2 radicals that would oxidize any organic compounds at the surface.[53] A record of organics on Mars could conceivably be extant, however, below the surface. On Venus, despite the presence of an ultraviolet shield, it is unlikely that ancient organics, if they ever existed, could be recoverable from the surface since it is both very hot (743 K) and relatively young (<500 million years).

Abiotic Organic Synthesis in the Interior of Earth

There is a broad range of physical environments across the surfaces and in the interiors of the terrestrial planets; some of these may support conditions suitable for the abiotic synthesis of organic material. Notably, understanding of abiotic synthesis within the interior of Earth has grown considerably.

Organic material has been synthesized abiotically throughout Earth's history. For example, abiotic methane has been detected in fluids emitted in deep-sea hydrothermal vents,[54] as fluid inclusions within recently formed ocean-crustal rocks (mid-oceanic ridge basalts),[55] and in remarkably preserved ancient seafloor rocks some 3.2 billion years old.[56] Locally significant quantities of hydrocarbon gases (methane, ethane, and propane) have been generated and reservoired in ancient volcanogenic massive sulfide deposits (e.g., the 2.7 billion-year-old Kidd Creek deposit in Ontario).[57]

There are two predominant sources for abiotic methane in the interior of Earth:

- The direct formation from carbon in volatile-rich fluids associated with the partial melting of rocks deep within Earth's interior; and
- Formation in the vicinity of mid-oceanic ridges or spreading centers.

These two possibilities are discussed in detail in the next two sections.

Methane Formation in Earth's Deep Interior

Methane may form directly from carbon in volatile-rich fluids derived from partial melting of rocks deep within Earth's interior. Carbon in such fluids is found in a number of forms/species, including CH_4, CO_2, CO, and graphite. At very high temperatures (e.g., 1300 K), thermodynamic calculations indicate that CO_2 and CO should be the dominant, carbon-bearing phases for fluids, with CH_4 essentially absent.[58] Analyses of high-temperature volcanic gases (900 to 1500 K) support this conclusion, revealing no or only trace quantities of CH_4.[59] At lower temperatures, however, thermodynamic calculations predict that CH_4 should be predominant.[60] Shock proposed that if kinetic barriers suppressed the equilibrium shift from a predominance of CO_2 and H_2 to CH_4 and H_2O, reduction of CO_2 to form other hydrocarbons would still be thermodynamically favorable (in the absence of appropriate catalysts, the methane-forming reaction is kinetically inhibited at low temperatures).[61]

Methane Formation at Mid-Ocean Ridges or Spreading Centers

One of the primary mechanisms by which Earth loses heat is through the generation of new oceanic crust. This crust is basaltic and is created by the partial melting of mantle rocks deep within Earth, leading to melt migration through the mantle and its emergence along various mid-oceanic spreading centers deep under the oceans. While considerable heat is lost immediately at the spreading center, sufficient remains in the new seafloor to initiate hydrothermal circulation cells on the flanks of a spreading center whereby cold ocean waters are drawn down and circulate in freshly fractured hot seafloor.[62] Once these fluids reach sufficient temperatures, they are capable of reacting with the new oceanic crust before they rise, completing the circulation pathway. If recirculating hydrothermal fluids contain dissolved CO_2, then methane can be formed. This has been shown to occur experimentally.[63] In natural, deep-ocean systems, methane plumes have been detected originating from deep-sea spreading centers as well as rift-valley regions in their vicinity.[64] Whereas in some cases this CH_4 is clearly biological, in many cases its relatively heavy carbon isotopic composition suggests abiological synthesis. The CH_4 formation mechanism is as follows. The hydrothermal alteration (serpentinization) of basalt and exposed peridotite (in particular, transformation of the major mineral constituent, olivine) yields a more magnesium-rich hydrated silicate, serpentine. The excess ferrous iron reacts with water to produce magnetite (mixed ferrous and ferric iron) and hydrogen gas. It is likely that this newly formed magnetite provides a suitable catalyst for methane formation. Furthermore, in experiments designed to replicate this chemistry, ethane and propane also formed in proportions $CH_4 > C_2H_6 > C_3H_8$.

A proposed physical scenario for such chemistry in natural systems is as follows. Fluids containing CO_2 and H_2 originate at depth within the new oceanic crust via the previously described processes. The fluids migrate upward to the surface via a fracture-pore network. The fluids pass through rock containing catalytic mineral phases. These catalysts promote the dissociation of CO_2 and drive catalytic hydrogenation of carbon, leading to the synthesis of CH_4. However, in addition to methane, sequential insertion of CO followed by reduction leads to chain growth and the formation of ethane, propane, and higher homologs.[65] These types of reactions are grouped under the umbrella of Fischer-Tropsch (FT) syntheses.[66] These reactions have also been shown to occur experimentally, producing not only hydrocarbons but also fatty acids, fatty alcohols, and other compounds.[67,68] The exact distribution in chain lengths and compounds is a complex function of fluid composition, temperature, pressure, the nature of the catalyst, and residence time.

Given this reasonable scenario, it is not surprising that FT chemistry may occur naturally in hydrothermal vent systems as well as in volcanogenic massive sulfide deposits. As for the significance of such systems as global sources of abiotic organic carbon, it has been estimated that between 10^8 and 10^9 kg of organic material per year could be synthesized via hydrothermal systems.[69] Placed in context with the major source of organic carbon synthesis on present-day Earth, primary productivity (i.e., the conversion of CO_2 to biomass via photosynthesis) yields approximately 5×10^{13} kg of organic carbon per year.[70] Thus, while hydrothermal sources of organic carbon may be significant, they are considerably less prolific on a planetary scale than biological organosynthesis, but may have been important as sources of prebiotic organic matter on the early Earth.[71]

Abiotic Organic Synthesis in the Interiors of Mars, Venus, and Mercury

Given that organic materials are being synthesized abiotically on Earth, it is possible to speculate on the extent to which this might have occurred or is currently occurring on the other terrestrial planets. The likelihood of a thermal gradient across the accretionary disk[72] suggests that the amount of hydrogen (predominantly in the form of water) and carbon (likely as CO_2 or CO_3^{2-}) remaining in the dust grains would decrease moving inward from Mars, past Earth and Venus, and ultimately to Mercury. However, this simple picture assumes that the protoplanetary dust particles and the progressively larger bodies that accreted from them all derived from the immediate neighborhood of the growing planets. Recent work has shown that considerable radial migration of source material due to collision and scattering of protoplanetary nuclei was likely, perhaps enhanced by the perturbative effects of an early-formed Jupiter. As a result, more-volatile-rich, cooler material might have contributed to the terrestrial planets.[73]

Taking these factors into account, one can speculate on the probability of abiotic organosynthesis on the terrestrial planets other than Earth. As stated earlier, Mars and Venus are likely to have formed from material that was compositionally similar to that of Earth. This similarity would suggest that all three planets started out with comparable inventories of water and carbon. Both Mars and Venus exhibit evidence of extensive volcanism. The estimated composition of their mantles suggest that, at least early in their respective histories, both planets had the potential for generation of at least methane and hydrocarbons, either via the thermal equilibration of carbon-bearing volcanic exhalations or via hydrothermal production if extensive bodies of water once existed.

The predominant gas species in the atmospheres of both Mars and Venus are CO_2 and smaller amounts of N_2 and CO. Such compositions are consistent with each planet's atmosphere being formed from high-temperature volcanic exhalation derived from partial melting of mantle rocks.

On Venus, the high surface temperatures set boundary conditions that might suppress the production of methane from sources other than direct volcanic exhalation. Lacking equilibration to form methane in the cooler regions of Venus's upper atmosphere, methane generation may not therefore occur anywhere on Venus. On Mars, however, formation and trapping of methane might be possible. Indeed, interest in this possibility has been greatly enhanced by claims of the spectroscopic detection of methane in the planet's atmosphere both by ground-based telescopes[74,75] and by the Mars Express spacecraft.[76] Although the results obtained from Mars Express are highly controversial, all three sets of observations indicate methane at concentrations of about 10 parts per billion. This is significant in that methane is unstable in the martian atmosphere and would disappear in ~300 years if not replenished. Although the origin of the methane has not yet been determined, possible sources include volcanic activity, chemical reactions between water and iron-bearing minerals in a hydrothermal system, and biological activity. Lower-temperature exhalations, e.g., on the flanks of the great martian shield volcanoes, could have equilibrated to form methane from CO_2 and H_2 in the relatively cool regolith and shallow lithosphere. If so, such methane might accumulate in the subsurface as stable methane hydrates.[77]

No surface water currently exists on Venus or Mars. Consequently, the generation of methane and higher-molecular-weight, organic matter via hydrothermal processes is currently not viable. In the case of Mars, however, there is strong evidence from the Opportunity and Spirit rovers that there was surface water early in martian history. Other missions have shown that there may have been lesser amounts of liquid water in recent history and that there is currently a large inventory of ground ice. This evidence, coupled with a history of active volcanism, perhaps as recently as ~10 Ma ago, suggests that extensive hydrothermal abiotic organic synthesis could have occurred. The ideal mode of preservation of such carbon would be as organic molecules trapped in fluid inclusions in rocks that constituted ancient martian seafloor or in silicified deposits in proximity to previously active hot springs.

The preservation of organic carbon in shallow martian regolith may be problematic. On Mars, the ultraviolet flux is sufficiently high that photolysis of water entrained within the surface sediment may provide a source of OH radicals. These radicals may attack organic matter directly or recombine to form hydrogen peroxide that will oxidize organic carbon upon contact. Both these oxidants will be found predominantly in the photic zone.[78] Organic carbon below the surface may also be threatened, not by direct reaction with OH radicals or peroxide but by reaction with strong inorganic oxidants (e.g., ferrate salts) that may mix down through regolith during wind-assisted sorting.

Reaction of any of these strong oxidants in the martian soil with organic matter will either destroy the organic matter totally or oxidize it partially to form products such as polycarboxylated aromatic acids.[79] Under high-soil-pH conditions, these molecules would reside as salts in the martian regolith. Thus, such acids, even if present in significant abundance, would not be volatile enough to allow detection by the instrumentation on board the Viking lander. If this scenario has merit, then preservation of organic matter in the martian shallow subsurface may be minimal, even though substantial reservoirs of abiotic organic matter might still exist deeper into the martian surface. There is the strong possibility of water ice a meter below the martian surface.

Finally, there is the special case of Mercury. Several arguments suggest that abiotic organic synthesis within Mercury's interior was and is insignificant. First, Mercury is considered a refractory planet in the sense that it presumably accreted from the most devolatilized dust grains. Thus, its carbon and hydrogen budget would have been low. Even if Mercury had started with carbon and H_2O contents proportionally similar to those of Venus and Earth, the proposed, late enormous collision stripped off much of the outer, silicate-rich crust and, with it, the majority of volatile material.[80] Second, although little is known regarding the tectonic history of Mercury, flyby imaging has not revealed any evidence of extensive volcanism on the surface of Mercury. Thus extensive exhalation of volatiles is not expected.

TERRESTRIAL PLANETS: RECOMMENDATIONS

Earth

Abiotic sources of organic carbon on Earth are only now receiving attention. Over the past two decades, the pace of research focusing on the organic, inorganic, and biochemistry of deep-sea hydrothermal vents has been increasing. Notwithstanding these advances, little is known regarding the range of organic compounds synthesized abiologically in these deep vent systems. Much of the recent exploration has focused on marine biology in close proximity to "black smokers." Although these black smokers are spectacular, the lower-temperature regimes on the flanks of oceanic spreading centers may constitute more important and prolific environments from the perspective of the abiological synthesis of organic carbon. Several projects are in progress. They include the following:

- The RIDGE (Ridge Interdisciplinary Global Experiments) program, a National Science Foundation initiative that lasted for 12 years starting in 1988, promoted interdisciplinary study, scientific communication, and outreach related to all aspects of the globe-encircling, mid-ocean ridge system. Research derived from this program has provided the vast majority of information regarding the biology, geology, and geophysics of these remarkable deep-sea environments. A continuation of the RIDGE program, RIDGE 2000, was created with the input of more than 200 U.S. scientists and is funded by the National Science Foundation. Compared to its predecessor, RIDGE 2000 has a more focused range of scientific priorities, with a greater emphasis on hydrothermal ecosystems and on understanding those systems in the context of regional and local volcanic and tectonic characteristics of specific sites.
- The North East Pacific Time-series Undersea Networked Experiments (Neptune) project is designed to create permanent undersea laboratories, initially along the Juan de Fuca ridge (a spreading center off the northwest coast of North America) and later at numerous other hydrothermal sites, to continuously monitor the chemistry, biology, and geology of these poorly understood terrains.[81] The goal of the Neptune project is to establish a regional ocean observatory in the northeast Pacific Ocean. The project's 3000-km network of fiber-optic/power cables will crisscross the Juan de Fuca region. Shore-based researchers will be able to interact with and obtain real-time data from instruments at and between these sites and thus monitor the physical, chemical, and biological phenomena taking place across several hundred thousand square kilometers of seafloor. Planning for Neptune is supported by the Keck Foundation and several federal agencies, including the National Science Foundation. Canadian participation in the Neptune project is funded by the Canadian Foundation for Innovation and the British Columbia Development Fund.
- The Archean Park program, an interdisciplinary characterization of the subseafloor biosphere, was initiated in 2000 and is led by scientists at the University of Tokyo. In 2001, the project drilled into the Suiyo Seamount (depth = 1380 m), a hydrothermally active volcano of the Izu-Bonin Arc (an active spreading center) in the Philippine Sea, and retrieved samples for biochemical and geochemical analyses. The goal of the program is to identify deep hydrothermal ecosystems where continental influence has been shown to be nonexistent.

While intense interest exists within the scientific community regarding the potential role of deep-sea hydrothermal vents in the origins of life and Earth's early atmosphere, the difficulty of studying the reaction zones in these remarkable natural sources of organic carbon precludes rapid development of a thorough physicochemical and biological understanding of such environments. Augmenting these studies are investigations examining the preserved vestiges of ancient spreading centers distributed throughout the geological record. Due to unusual and fortuitous aspects of deformation associated with plate collisions resulting from continental tectonics, sections of ancient seafloor now constitute coastal mountain ranges at numerous locations across the globe. Similarly, rocks associated with ancient spreading centers are preserved in ~3.3 billion-year old sediments of Western Australia and the 3.2 billion-year-old strata of South Africa. Studies of the inorganic chemistry of these rocks as well as analysis of any organic molecules trapped within fluid inclusions will go a long way toward improving understanding of the potential scope of abiological organosynthesis on early Earth and by comparison toward perhaps providing insight relevant to early Mars and Venus as well.

Mars

There are numerous reasons to expect that Mars should contain a rich inventory of organic carbon. However, past missions have shown that this carbon is not easily detected in the martian regolith and is, thus far, not detectable remotely. The principal problem may be that strong oxidants are formed in the shallow regions of the martian soil. Awareness of this potential problem arose from the results of the soil analyses made by the 1976 Viking landers. NASA's next Mars mission, Phoenix, is instrumented for the analysis of volatile organics released on heating martian soils collected in the northern polar region of the planet. A more thorough search for martian organics will be undertaken by NASA's Mars Science Laboratory, scheduled for launch in 2009.

The recognition of the martian meteorites has significantly expanded knowledge of the chemistry, mineralogy, age, and isotopic composition of the environments in martian crust. However, most of the meteorites in hand are young, are igneous in nature, and appear to be from a similar location. As several studies have now indicated, none of these rock types appear to be conducive to the preservation and accumulation of organic matter. In formulating a strategy to search for organic environments on Mars and elsewhere in the solar system, researchers can draw some insight from Earth's geologic record.

On Earth, the most suitable lithologies for the preservation and accumulation of organic matter are sedimentary rocks that are typically fine-grained and are characterized by well-defined aqueously derived mineral assemblages. Thus, it may be possible to obtain additional information about the associated organic matter present in these martian mineral assemblages in a single measurement of the organic and inorganic material present.

The detection and analysis of organic carbon on Mars will not be a simple task. There are likely to have been both exogenous and endogenous sources, the latter possibly carrying a biological signature. Furthermore, it is possible that life arose early in martian history, but may have not survived; thus the same difficult issues regarding identifying biomarkers arise for Mars as they do in studies of the most ancient rocks on Earth. Two aspects are then critical when prospecting for organic carbon on Mars. First, it is necessary to be able to distinguish between both endogenous and exogenous organics. Second, it is necessary to distinguish between organics derived from biotic and abiotic sources. From the perspective of identifying endogenous sources of organic carbon, therefore, it is critical that samples be obtained and analyzed from terrains where the formation history is well understood. In light of recent discoveries of surface water and volcanism in Mars's relatively recent past, the existence of a small remnant martian biosphere cannot be excluded.

Recommendation: Currently planned missions to Mars should seek to identify silicified martian terrains associated with ancient low-temperature hot springs in concert with a high probability of ground ice deposits to locate organic materials formed on Mars. Similarly, the identification of shallow marine and/or lacustrine sediments would provide another terrain well worth exploring in future missions as sites for martian endogenous organosynthesis.

It has been proposed that any organic matter in the martian regolith, either endogenous or exogenous, will have been modified via reaction with strong oxidants present in the soil. Carefully designed laboratory experiments will allow an assessment of this problem and will point to the most effective strategies for direct analysis of organic materials by future Mars landers. For example, chemical derivatization schemes would be necessary to produce volatile organic compounds for gas chromatographic analysis if the only surviving organic molecules found in the regolith are polycarboxylated aromatics.

Can a suitable simulated martian regolith be devised to support such studies? There is information on the elemental composition of regolith from the Viking landers, the Mars Pathfinder mission, and the Mars Exploration Rovers. In addition, there are analytical results from the SNC meteorites that,[82] together with spectroscopic data, provide insights into the regolith's composition. On the basis of these data, several analogs of martian regolith have been prepared.[83] Although the oxidants in the soil have not been identified, it is possible that they are peroxides that were generated by short-wavelength ultraviolet light. Oxidation studies of organics using various regolith analogs and a variety of possible oxidants will provide insight into the likely oxidation products that will be observed on Mars. This information will be crucial to the design of the analytical capabilities of the lander that will perform the analyses of the oxidation products on Mars. Experiments designed to reproduce the chemical characteristics of martian regolith will lead to development of analytical strategies with the broadest bandwidth for detection of organic molecules. Assessment of the time scales for oxidative alteration of organic materials in the martian regolith would address issues related to optimal minimum drilling depths for future Mars lander missions.

Recommendation: Laboratory models of Mars soil chemistry should be used to study plausible mechanisms for the oxidative alteration of organic materials in the martian regolith and to evaluate their integrated effects. Materials studied should include likely exogenous products (organic compounds like those found in meteorites) as well as conceivable martian pre- and postbiotic products.

As instrument development continues for future robotic missions to Mars, it is important that such missions be capable of assessing as fully as possible the inventory of organic matter there. Clearly such development should be strongly guided by the information provided by the Spirit and Opportunity discoveries.

Although such instruments will likely include mass spectrometers, considerable effort is required in the sample preparation stages prior to mass spectrometric analysis. Specifically, given what is now known about the regolith at the Spirit landing site and the outcrops at the Opportunity landing site, questions arise as to how to best isolate organic matter from the inorganic matrix and how to best introduce this organic matter into the source of a mass spectrometer.

Even though future robotic missions will be equipped with instrumentation to analyze samples (e.g., the Mars Science Laboratory), these analyses will never be able to achieve the capabilities of Earth-bound laboratories. The discoveries by the rover Opportunity of what appears to be an unambiguously sedimentary outcrop greatly increases the impetus for martian sample return missions. Similarly, the discovery of the halogens bromine and chlorine in abundance at the location of the Spirit rover landing site strongly suggests the former presence of surface water. Samples from either location might very well contain organic matter derived from extinct (or perhaps even extant) life. The successes of Spirit and Opportunity further illustrate the importance of planning future missions to bring martian samples back to Earth.

NOTES

1. W.J. Schopf and B.M. Packer, "Early Archean (3.3-Billion-to-3.5-Billion-Year-Old) Microfossils from the Warrawoona Group, Australia," *Science* 237: 70-73, 1987.

2. M. Schidlowski, "A 3,800-Million Year Isotopic Record of Life from Carbon in Sedimentary Rocks," *Nature* 333: 313-318, 1988.

3. K.A. Maher and D.J. Stevenson, "Impact Frustration of the Origin of Life," *Nature* 331: 612-614, 1988.

4. M. Schidlowski, "A 3,800-Million Year Isotopic Record of Life from Carbon in Sedimentary Rocks," *Nature* 333: 313-318, 1998.

5. S.J. Mojzisis, G. Arrhenius, K.D. McKeegan, T.M. Harrison, A.P. Nutman, and C.R.L. Friend, "Evidence for Life on Earth Before 3,800 Million Years Ago," *Nature* 384: 55-59, 1996.

6. C. Chyba, "The Heavy Bombardment and the Origins of Life," *Astronomy* 20(11): 28-35, 1992.

7. See, for example, A.L. Burlingame, M. Calvin, J. Han, W. Henderson, W. Reed, and B.R. Simoneit, "Lunar Organic Compounds: Search and Characterization," *Science* 167: 751-752, 1970.

8. K. Watson, B. Murray, and H. Brown, "The Behavior of Volatiles on the Lunar Surface," *Journal of Geophysical Research* 66: 3033-3045, 1961.

9. J.R. Arnold, "Ice in the Lunar Polar Regions," *Journal of Geophysical Research* 84: 5659-5668, 1979.

10. W.C. Feldman, S. Maurice, A. Binder, B.L. Barraclough, R.C. Elphic, and D.J. Lawrence, "Fluxes of Fast and Epithermal Neutrons from Lunar Prospector: Evidence for Water Ice at the Lunar Poles," *Science* 281: 1496-1500, 1998.

11. P.G. Lucey, "Potential for Prebiotic Chemistry at the Poles of the Moon," pp. 84-88 in *Instruments, Methods, and Missions for Astrobiology III* (R.B. Hoover, ed.), Proceedings of SPIE, Vol. 4137, 2000.

12. National Research Council, *New Frontiers in the Solar System: An Integrated Exploration Strategy*, The National Academies Press, 2003, pp. 4-6 and 194.

13. National Research Council, *New Frontiers in the Solar System: An Integrated Exploration Strategy*, The National Academies Press, 2003, pp. 182-184.

14. G.W. Wetherill, "Provenance of the Terrestrial Planets," *Geochimica et Cosmochemica Acta* 58: 4513-4520, 1994.

15. R.H. Whittaker and G. Likens, "Carbon in the Biota," *Carbon in the Biosphere: Proceedings of the 24th Brookhaven Symposium in Biology* (G.M. Woodwell and E.V. Pecan, eds.), United States Atomic Energy Commission, 1973. Available from National Technical Information Service, Springfield, Va.

16. D.H. Welte, "Organischer Kohlenstoff und die Entwicklung der Photosynthese auf der Erde," *Naturwissenschaften* 57: 17-23, 1970.

17. D.W. Van Krevelen, *Properties of Polymers*, 3rd Edition, Elsevier Science, Amsterdam, The Netherlands, 1990.

18. J.M. Hunt, *Petroleum Geochemistry and Geology*, 2nd edition, W.H. Freeman and Co., New York, 1996, pp. 18-21.

19. C.F. Chyba and C. Sagan, "Endogenous Production, Exogenous Delivery, and Impact-shock Synthesis of Organic Molecules: An Inventory for the Origins of Life," *Nature* 355: 125-132, 1992.

20. J. Oró, "Comets and the Formation of Biochemical Compounds on the Primitive Earth," *Nature* 190: 389-390, 1961.

21. C.F. Chyba and G.D. McDonald, "The Origin of Life in the Solar System: Current Issues," *Annual Review of Earth and Planetary Sciences* 23: 215, 1995.

22. C. Sagan and C. Chyba, "The Early Faint Sun Paradox: Organic Shielding of Ultraviolet-labile Greenhouse Gases," *Science* 276: 1217-1221, 1997.

23. S.L. Miller and L.E. Orgel, *The Origins of Life on the Earth*, Prentice Hall, Englewood Cliffs, N.J., 1974.

24. T.M. McCollom and J.S. Seewald, "Experimental Constraints on the Hydrothermal Reactivity of Organic Acids and Acid Anions: I. Formic Acid and Formate," *Geochimica et Cosmochimica Acta* 67: 3625-3644, 2003.

25. T.M. McCollom and J.S. Seewald, "Abiotic Synthesis of Organic Compounds in Deep-sea Hydrothermal Environments," *Chemical Reviews* (submitted).

26. B. Sherwood Lollar, T.D. Westgate, J.A. Ward, G.F. Slater, and G. Lacrampe-Couloume, "Abiogenic Formation of Alkanes in the Earth's Crust as a Minor Source for Global Hydrocarbon Reservoirs," *Nature* 416: 522-524, 2002.

27. T.M. McCollom, G. Ritter, and B.R. Simoneit, "Lipid Synthesis Under Hydrothermal Conditions by Fischer-Tropsch Type Reactions," *Origins of Life and Evolution of the Biosphere* 29: 153-166, 1999.

28. G.D. Cody, N.Z. Boctor, R.M. Hazen, J.A. Brandes, H.J. Morowitz, and H.S. Yoder, Jr., "Geochemical Roots of Autotrophic Carbon Fixation: Hydrothermal Experiments in the System Citric Acid, H_2O-($\pm FeS$)-($\pm NiS$)," *Geochimica et Cosmochimica Acta* 65: 3557-3576, 2001.

29. A.I. Rushdi and B.R.T. Simoneit, "Condensation Reactions and Formation of Amides, Esters and Nitriles Under Hydrothermal Conditions," *Astrobiology* 4: 211-224, 2004.

30. E. Shock, "Chemical Environments of Submarine Hydrothermal Systems," *Origin of Life and Evolution of the Biosphere* 22: 67-107, 1992.

31. J.L. Charlou et al., "Intense CH_4 Plumes Generated by Serpentinization of Ultramafic Rocks at the Intersection of the 15°20' Fracture Zone and the Mid-Atlantic Ridge," *Geochimica et Cosmochimica Acta* 62: 2323-2333, 1998.

32. G. Flynn, "The Delivery of Organic Matter from Asteroids and Comets to the Early Surface of Mars," *Earth, Moon, and Planets* 72: 469-474, 1996.

33. See, for example, S.A. Benner, K.G. Devine, L.N. Matveeva, and D.H. Powell, "The Missing Organic Molecules on Mars," *Proceedings of the National Academy of Sciences* 97: 2425-2430, 2000.

34. M.J. Mumma, R.E. Novak, M.A. DiSanti, and B.P. Bonev, "A Sensitive Search for Methane on Mars," AAS/DPS 35th Meeting, September 1-6, 2003.

35. V.A. Krasnopolsky, J.P. Maillard, and T.C. Owen, "Detection of Methane in the Martian Atmosphere: Evidence for Life," European Geophysical Union Meeting, Nice, May 2004.

36. V. Formisano, S. Atreya, T. Encrenaz, N. Ignatiev, and M. Giuranna, "Detection of Methane in the Atmosphere of Mars," *Science* 306: 1758-1761, 2004.

37. A.A. Pavlov, J.F. Kasting, L.L. Brown, K.A. Rages, and R. Freedman, "Greenhouse Warming by CH_4 in the Atmosphere of Early Earth," *Journal of Geophysical Research—Planets* 105: 11981-11990, 2000.

38. C. Sagan and C. Chyba, "The Early Faint Sun Paradox: Organic Shielding of Ultraviolet-labile Greenhouse Gases," *Science* 276: 1217-1221, 1997.

39. J.W. Delano, "Redox History of the Earth's Interior Since ~3900 Ma: Implications for Prebiotic Molecules," *Origins of Life and Evolution of the Biosphere* 31: 311-341, 2001.

40. S.L. Miller, "Production of Amino Acids Under Possible Primitive Earth Conditions," *Science* 117: 528-529, 1953.

41. S.L. Miller, "The Endogenous Synthesis of Organic Compounds," pp. 59-85 in *The Molecular Origins of Life* (A. Brack, ed.), Cambridge University Press, Cambridge, England, 1998.

42. C. Sagan and C. Chyba, "The Early Faint Sun Paradox: Organic Shielding of Ultraviolet-labile Greenhouse Gases," *Science* 276: 1217-1221, 1997.

43. J.F. Kasting and L.L. Brown, "The Early Atmosphere as a Source of Biogenic Compounds," pp. 35-56 in *The Molecular Origins of Life* (A. Brack, ed.), Cambridge University Press, Cambridge, England, 1998.

44. For a review of relevant issues see, for example, J. Oró, S.L. Miller, and A. Lazcano, "The Origin and Early Evolution of Life on Earth," *Annual Reviews of Earth and Planetary Sciences* 18: 317-356, 1990.

45. K.J. Zanhle, "Photochemistry of Methane and the Formation of Hydrocyanic Acid (HCN) in the Earth's Early Atmosphere," *Journal of Geophysical Research* 91: 2819-2834, 1986.

46. J.P. Ferris, P.D. Joshi, E.H. Edelson, and J.G. Lawless, "HCN: A Plausible Source of Purines, Pyrimidines and Amino Acids on the Primitive Earth," *Journal of Molecular Evolution* 11: 293-311, 1978.

47. J.P. Pinto, C.R. Gladstone, and Y.L. Yung, "Photochemical Production of Formaldehyde in the Earth's Primitive Atmosphere," *Science* 210: 183-185, 1980.

48. J.P. Pinto, C.R. Gladstone, and Y.L. Yung, "Photochemical Production of Formaldehyde in the Earth's Primitive Atmosphere," *Science* 210: 183-185, 1980.

49. Y.L. Yung, and W.B. DeMore, *Photochemistry of Planetary Atmospheres*, Oxford University Press, New York, 1999.

50. D.H. Grinspoon, *Venus Revealed: A New Look Below the Clouds of Our Mysterious Twin Planet*, Perseus Publishing, Cambridge, Mass., 1997.

51. D. Schulze-Makuch and L.N. Irwin, "Reassessing the Possibility of Life on Venus: Proposal for an Astrobiology Mission," *Astrobiology* 2: 197-202, 2002.

52. D. Schulze-Makuch, O. Abbas, L.N. Irwin, and D.H. Grinspoon, "Microbial Adaption Strategies for Life in the Venusian Atmosphere," Abstract 12747, NASA Astrobiology Institute General Meeting, 2003.

53. J.P. Ferris, "Photochemical Transformations on the Primitive Earth and Other Planets," Chapter 1 of *Organic Photochemistry*, Volume 8 (A. Padwa, ed.), Marcel Dekker, New York, 1987.

54. J.L. Charlou, Y. Fouquet, H. Bougault, J.P. Donval, J. Etoubleau, P. Jean-Baptiste, A. Dapoigny, P. Appriou, and P.A. Rona, "Intense CH_4 Plumes Generated by Serpentinization of Ultramafic Rocks at the Intersection of the $15°20'N$ Fracture Zone and the Mid-Atlantic Ridge," *Geochimica et Cosmochimica Acta* 62: 2323-2333, 1998.

55. D.S. Kelly, "Methane-rich Fluids in the Oceanic Crust," *Journal of Geophysical Research* 101: 2943-2962, 1996.

56. C.E.J. de Ronde, D.E. Der Channer, K. Faure, C.J. Bray, and E.T.C. Spooner, "Fluid Chemistry of Archean Sea-floor Hydrothermal Vents: Implications for the Composition of Circa 3.2 Ga Seawater," *Geochimica et Cosmochimica Acta* 61: 4025-4042, 1997.

57. B. Sherwood Lollar, T.D. Westgate, J.A. Ward, G.F. Slater, and G. Lacrampe-Couloume, "Abiogenic Formation of Alkanes in the Earth's Crust as a Minor Source for Global Hydrocarbon Reservoirs," *Nature* 416(6880): 522-524, 2002.

58. H.P. Eugster and G.B. Skippen, "Igneous and Metamorphic Reactions Involving Gas Equilibria," *Researches in Geochemistry*, Volume 2 (P.H. Abelson, ed.), John Wiley and Sons, New York, 1967.

59. R.B. Symonds, W.I. Rose, D.J.S. Bluth, and T.M. Gerlach, "Volcanic Gas Studies: Methods, Results, and Applications," *Volatiles in Magmas* (M.R. Carroll and J.R. Holloway, eds.), *Reviews in Mineralogy* 30: 1-60, 1994.

60. H.P. Eugster and G.B. Skippen, "Igneous and Metamorphic Reactions Involving Gas Equilibria," *Researches in Geochemistry*, Volume 2 (P.H. Abelson, ed.), John Wiley and Sons, New York, 1967.

61. E.L. Shock, "Geochemical Constraints on the Origin of Organic Compounds in Hydrothermal Systems," *Origins of Life and Evolution of the Biosphere* 20: 331-367, 1990.

62. N.G. Holm and E.M. Andersson, "Hydrothermal Systems," pp. 86-99 in *The Molecular Origins of Life* (A. Brack, ed.), Cambridge University Press, Cambridge, England, 1998.

63. M.E. Berndt, D. Allen, and W.E. Seeyfried, "Reduction of CO_2 During Serpentinization of Olivine at 300°C and 500 Bar," *Geology* 24: 351-354, 1996.

64. J.L. Charlou, Y. Fouquet, H. Bougault, J.P. Donval, J. Etoubleau, P. Jean-Baptiste, A. Dapoigny, P. Appriou, and P.A. Rona, "Intense CH_4 Plumes Generated by Serpentinization of Ultramafic Rocks at the Intersection of the $15°20'N$ Fracture Zone and the Mid-Atlantic Ridge," *Geochimica et Cosmochimica Acta* 62: 2323-2333, 1998.

65. H. Koch and W. Gilfert, "Carbonsäure Synthese aus Olefinen, Kohlenoxyd und Wasser," *Brennstoff-Chemie* 36: 321-352, 1955.

66. F. Fischer, "Synthese der Treibstoffe (Kogasin) und Schmieröle aus Kohlenoxyd und Wasserstoff bei Gewöhnlichem Druck," *Brennstoff-Chemie* 16: 1-11, 1935.

67. T.M. McCollom, G. Ritter, and B.R. Simoneit, "Lipid Synthesis Under Hydrothermal Conditions by Fischer-Tropsch Type Reactions," *Origins of Life and Evolution of the Biosphere* 29: 153-166, 1999.

68. A.I. Rushdi and B.R.T. Simoneit, "Lipid Formation by Aqueous Fischer-Tropsch-type Synthesis over a Temperature Range of 100-400°C," *Origins of Life and Evolution of the Biosphere* 31: 103-118, 2001.

69. E. Shock, "Chemical Environments of Submarine Hydrothermal Systems," *Origins of Life and Evolution of the Biosphere* 22: 67-107, 1992.

70. R.H. Whittaker and G. Likens, "Carbon in the Biota," *Carbon in the Biosphere: Proceedings of the 24th Brookhaven Symposium in Biology* (G.M. Woodwell and E.V. Pecan, eds.), United States Atomic Energy Commission, 1973. Available from National Technical Information Service, Springfield, Va.

71. N.G. Holm, ed., "Marine Hydrothermal Systems and the Origin of Life," *Origins of Life and Evolution of the Biosphere* 22: 1-242, 1992.

72. A. Boss, "Evolution of the Solar Nebula: I. Nonaxisymmetric Structure During Nebula Formation," *Astrophysical Journal* 345: 544-571, 1989.

73. T.C. Owen and A. Bar-Nun, "Volatile Contributions from Icy Planetesimals," pp. 459-471 in *Origin of the Earth and Moon* (R.M. Canup and K. Righter, eds.), University of Arizona Press, Tucson, Ariz., 2000.

74. M.J. Mumma, R.E. Novak, M.A. DiSanti, and B.P. Bonev, "A Sensitive Search for Methane on Mars," AAS/DPS 35th Meeting, September 1-6, 2003.

75. V.A. Krasnopolsky, J.P. Maillard, and T.C. Owen, "Detection of Methane in the Martian Atmosphere: Evidence for Life," European Geophysical Union Meeting, Nice, May 2004.

76. V. Formisano, S. Atreya, T. Encrenaz, N. Ignatiev, and M. Giuranna, "Detection of Methane in the Atmosphere of Mars," *Science* 306: 1758-1761, 2004.

77. I-M. Chou, A. Sharma, R.C. Burress, J. Shu, H-K. Mao, R.J. Hemley, A.F. Goncharov, L.A. Stern, and S.H. Kirby, "Transformations in Methane Hydrates," *Proceedings of the National Academy of Sciences* 97: 13484-13487, 2000.

78. S.A. Benner, K.G. Devine, L.N. Matveeva, and D.H. Powell, "The Missing Organic Molecules on Mars," *Proceedings of the National Academy of Sciences* 97: 2425-2430, 2000.

79. S.A. Benner, K.G. Devine, L.N. Matveeva, and D.H. Powell, "The Missing Organic Molecules on Mars," *Proceedings of the National Academy of Sciences* 97: 2425-2430, 2000.

80. G.W. Wetherhill, "Accumulation of Mercury from Planetismals," pp. 671-691 in *Mercury* (F. Vilas, C.R. Chapman, and M.S. Matthews, eds.), University of Arizona Press, Tucson, Ariz., 1988.

81. For general background on Neptune and related projects see, for example, Ocean Studies Board, National Research Council, *Illuminating the Hidden Planet: The Future of Seafloor Observatory Science*, National Academy Press, Washington, D.C., 2000.

82. See, for example, S.P. Kounaves, S.R. Lukow, B.P. Comeau, M.H. Hecht, S.M. Grannan-Feldman, K. Manatt, S.J. West, X. Wen, M. Frant, and T. Gillette, "Mars Surveyor Program '01 Mars Environmental Compatibility Assessment Wet Chemistry Lab: A Sensor Array for Chemical Analysis of the Martian Soil," *Journal of Geophysical Research* 108(E7): 5077, 2003; and M. Koel, M. Kaljurand, and C.H. Lochmuller, "Evolved Gas Analysis of Inorganic Materials Using Thermochromatography: Model Inorganic Salts and Palagonite Martian Soil Simulants," *Analytical Chemistry* 69: 4586-4591, 1997.

83. See, for example, T.L. Roush and J.B. Orenberg, "Estimated Detectability of Iron-Substituted Montmorillonite Clay on Mars from Thermal Emission Spectra of Clay-Palagonite Physical Mixtures," *Journal of Geophysical Research* 101(E7): 26111-26118, 1996; and M. Koel, M. Kaljurand, and C.H. Lochmuller, "Evolved Gas Analysis of Inorganic Materials Using Thermochromatography: Model Inorganic Salts and Palagonite Martian Soil Simulants," *Analytical Chemistry* 69: 4586-4591, 1997.

III—Exploration: Where to Go and What to Study

7

Approaches to Research

Intensive, programmatic studies of missions being directed to specific locations in the solar system can seek to identify the most efficient means of exploration, the pathway that would provide the greatest amount of decisive information in return for the simplest, quickest, and cheapest measurements and analyses. Such studies inevitably reflect diverse factors. The task group's purposes in this study are to focus on organic matter rather than on specific missions or locations, to call attention to objectives of particular importance, and to consider issues that might cut across otherwise-separate programs. To accomplish this, the task group set out to address the following questions:

1. What are the sources of reactants and energy that lead to abiotic synthesis of organic compounds and to their alteration in diverse solar system environments?
2. What are the distribution and history of reduced carbon compounds in the solar system, and which features of that distribution and history, or of the compounds themselves, can be used to discriminate among synthesis and alteration processes?
3. What are the criteria that distinguish abiotic from biotic organic compounds?
4. What aspects of the study of organic compounds in the solar system can be accomplished from ground-based studies (theoretical, laboratory, and astronomical), Earth orbit, and planetary missions (orbiters, landers, and sample return), and which new capabilities might have the greatest impact on each?

With these questions in mind, the task group sought to identify reservoirs of organic material in the solar system and to consider what is known about their history as well as their present composition. Two broad questions can be identified:

- What are the relationships between organic materials in diverse extraterrestrial settings such as planetary and satellite regoliths, asteroids, comets, and meteorites?
- What processes produced the organic materials?

Much can be inferred from compositional and isotopic data. For example, a particular set of organic compounds might be found in interstellar media. The same materials might occur in comets, and plausibly related materials could turn up in meteorites and on asteroids. The pattern would indicate a possible line of inheritance, showing that, at the time of its origin, the solar system incorporated interstellar organic material. The hypothesis could be reinforced or even confirmed decisively if the ratio of ^{13}C to ^{12}C were the same in all of the materials studied. A finding of shared origins would be profoundly significant.

The specific roster of identified compounds would be interesting, but the more general point that the organic chemistry of the solar system was connected to that of the broader cosmos would have many ramifications. If one set of compounds survived, what about others? If the solar system contained significant quantities of organic material from the outset, what consequences followed? Clearly, ongoing broad surveys of organic materials, particularly those that provide data that can establish relationships between diverse locales, should be encouraged. In such work, breadth, i.e., the examination of materials from the widest possible range of settings, could be as important as detail.

A second aspect of molecular history concerns synthetic processes. Wherever some organic material is found, and however it might be related to similar materials elsewhere, by exactly what process was it made? Were the atoms drawn from gaseous precursors? Did they combine on a surface? Can patterns of repetition (polymerization) be recognized? Were the chemical reactions highly selective, leading to only a few molecular structures, or was the range of products diverse? Were living organisms involved? For answering such questions, detail is essential. Investigators would like to determine precisely the structure, abundance, and isotopic composition of every compound. It would be better still to determine the distributions of isotopes within each compound (i.e., intramolecular patterns of isotopic order that could reveal how the components of the molecule were assembled). Since living organisms often utilize minerals in their metabolic processes, it will be important also to investigate the inorganic phases associated with the organic materials. Such thorough analysis will require samples large enough to sustain extensive dissection.

GENERAL STRATEGIES

Two general strategies for approaching both ground-based research and research carried out by flight missions are as follows:

Recommendation: Strategy 1—Every opportunity should be seized to increase the breadth and detail in inventories of organic material in the solar system. As results accumulate, each succeeding investigation should be structured to provide information that will allow improved comparisons between environments. Analyses should determine abundance ratios for the following:

- Compound classes (e.g., acetylenic, aliphatic, aromatic);
- Individual compounds (e.g., methane/ethane, HCN/HNC);
- Elements in organic material (e.g., C, H, N, O, S); and
- The isotopes of elements such as C, H, N, and O.

Investigators should strive to interpret these results in terms of precursor-product relationships.

These objectives are rudimentary compared to studies of, for example, the chirality of amino acids. They are, however, broadly applicable and represent systematic steps toward addressing questions of biogenicity, lines of inheritance of organic material, and mechanisms of synthesis. With limited funds, returns from investigations like those proposed as opportunities for research in Chapters 2 through 6 will move more smoothly toward ultimate success. For example, it is proposed that newer, more sensitive and specific analytical methods be used for the analysis and reanalysis of carbonaceous chondrites. As these studies proceed and the results from flight experiments are obtained, it will become apparent which of these new techniques should be adapted to flight experiments. Moreover, the ground-based investigations of chondrites will pave the way for better analyses of returned samples, whenever they become available.

Recommendation: Strategy 2—Organic-carbon-related flight objectives should be coordinated across missions and structured to provide a stepwise accumulation of basic results. Some of the objectives that should be included in such missions are as follows:

- Quantitation of the amount of organic carbon present to ± 30 percent precision and accuracy over a range of 0.1 ppm to 1 percent;
- Repetitive analyses of diverse samples at each landing site;
- Comparability so that relatable data are obtained from a wide range of sites; and
- Elemental and isotopic analyses so that the composition (H/C, N/C, O/C, and S/C) is obtained together with the isotope ratios of all the carbon-bearing phases.

These recommended approaches to research will allow scientists to build an overview of the distribution of organic carbon in the solar system; provide information about heterogeneity at each location studied; and support preliminary estimates of relationships, if any, between organic materials at diverse sites.

SELECTED OPPORTUNITIES FOR RESEARCH

Chapters 2 through 6 discuss a large number of research investigations that have the potential to significantly increase knowledge about the sources and history of carbon in the solar system. From those, the task group selected the research that seemed to promise the greatest return on investment. The selected research opportunities were then divided into three general categories based on the cost of the research and the time frame in which it could be carried out:

- Ground-based studies that can be carried out in the very near term and for a minimal cost relative to the other recommended research;
- Studies that can be carried out in the relatively near term—5 to 10 years—and are also supported by the findings and recommendations of the NRC's 2003 solar system exploration decadal strategy report,[1] which surveyed the broad community of scientists studying various aspects of the solar system, and, through a series of workshops and meetings, developed a roadmap of prioritized research for the next decade; and
- Far-term research recommended for its potential to expand knowledge of carbon compounds in the solar system and that would probably be carried out 10 years or more in the future but might require some near-term planning. This recommended research is ranked in terms of its potential for expanding knowledge of carbon compounds in the solar system and for its close relationship to research and missions currently in progress or recently completed.

The recommended research is presented in Chapters 2 through 6 and is summarized in the Executive Summary.

NOTE

1. National Research Council, *New Frontiers in the Solar System: An Integrated Exploration Strategy*, The National Academies Press, Washington, D.C., 2003.

Appendix

Glossary

Abiotic—Of or relating to nonliving things; independent of life or living organisms.

Accretion—The process by which an astronomical object increases in mass by the gravitational attraction of matter.

Albedo—The fraction of light that is reflected by a surface; commonly used in astronomy to describe the reflective properties of planets, satellites, and asteroids.

Amino acid—Any organic compound containing an amino ($-NH_2$) and a carboxyl ($-COOH$) group. There are 20 α-amino acids from which proteins are synthesized during ribosomal translation of mRNA.

Aromatic compound—A major class of unsaturated cyclic hydrocarbons characterized by the presence of one or more rings of carbon atoms. The class is typified by benzene, which has a six-carbon ring containing three pairs of alternating single and double bonds.

Aromaticity—A property of the arrangement of chemical bonds in cyclic hydrocarbons which confers enhanced stability to aromatic compounds.

Asteroid—Small, rocky bodies in orbit around the Sun, found mainly between the orbits of Mars and Jupiter.

Astronomical unit (AU)—The mean distance of Earth from the Sun.

Biotic—Of or relating to living things; caused or produced by living organisms.

C-, P-, and D-type asteroids—Classes of asteroids sorted according to their spectral reflectance. Asteroid classes are correlated with position within the asteroid belt, with increasingly red objects farther from the Sun. C-type asteroids have a mid-belt location, and P- or D-types are in the outer belt. C-types are a very good spectral match to carbonaceous chondrites; beyond 2.7 AU (the so-called soot line), asteroids are very carbon-rich (P and D types).

Carbonaceous chondrite—A rare type of stony meteorite that is rich in carbon compounds and is thought to be relatively unaltered since the beginning of the solar system. Its spectrum (and probably also its composition) closely resembles that of the C-type asteroids.

Catalyst—A substance that enhances the rate of reaction by providing a lower-energy alternative pathway.

Centaurs—A family of small solar system bodies found between the orbits of Jupiter and Neptune. Their orbital characteristics indicate that they have not resided in their present locations for very long, leading to the suggestion that they are recently migrated Kuiper belt objects and that further evolution of their orbits might turn them into short-period comets.

Chiral—Describing a molecule configured such that it cannot be superimposed on its mirror image.

Chromatography—A family of techniques for separating components from a mixture. All take advantage of the fact that different substances diffuse through a given medium at different rates. Gas chromatography is a technique for separating gas mixtures, in which the gas is passed through a long column containing a fixed absorbent phase that partitions the gas mixture into its component parts.

Coma—The quasispherical envelope of gas and dust surrounding the nucleus of an active comet, created when the ambient heat causes the sublimation of cometary ices.

Cosmic rays—High-energy charged particles consisting of atomic nuclei, electrons, and protons, which originate from the Sun and from energetic astrophysical processes (e.g., those associated with supernovas).

Desorption—A physical or chemical process by which a substance that has been adsorbed or absorbed by a liquid or solid material is removed from the material.

Diastereoisomers—Stereoisomers that are not mirror images of each other.

Diffuse interstellar bands (DIBs)—Unidentified absorption bands detected mostly in the visible spectrum of reddened O- and B-type stars and observed ubiquitously in space. Recent results point strongly toward a gas-phase molecular origin, but the DIB carriers' identification remains an extremely puzzling issue.

Enantiomer—Stereoisomers that are mirror images of each other. Enantiomers are optically active and rotate the plane of polarized light.

Exogenous delivery—Delivery of matter to a planetary environment via asteroidal or cometary impact.

Fischer-Tropsch (FT) catalytic process—A method for the synthesis of hydrocarbons and other aliphatic compounds. Typically, a mixture of hydrogen and carbon monoxide is reacted in the presence of an iron or cobalt catalyst to produce methane, synthetic gasoline, and other aliphatic compounds, with water and carbon dioxide as byproducts.

Free radical—A highly reactive chemical species carrying no charge and having a single unpaired electron in an orbital.

Fullerene—Any of various cagelike molecules that constitute the third form of pure carbon (along with the forms diamond and graphite), whose prototype C_{60} (buckyball) is the roundest molecule that exists. Fullerenes are a class of discrete molecules, soccerball-shaped forms of carbon with extraordinary stability (so named because their configuration suggests the shape of Buckminster Fuller's famous geodesic dome).

Fumarole—A vent, usually in volcanic regions, from which vapors or gases are released.

Geminate recombination—The reaction with each other of two transient molecular entities produced from a common precursor in solution.

Hale-Bopp (comet)—Correctly known as C/1995 O1 (Hale Bopp), it is the brightest comet to appear in the night sky for many decades. Discovered by Alan Hale and Thomas Bopp on July 22, 1995, it reached perihelion on April 1, 1997, and was visible to the naked eye for many months. Its nucleus appears to be very large, about 40 km across.

Heat pulse—A rapid change in ambient temperature conditions over a wide field of view. The temperature change can be higher or lower than that of the normal, ambient, slowly changing temperature.

Infrared—The portion of the electromagnetic spectrum radiation with wavelengths in the range from 750 nm to 0.1 mm.

Interstellar medium (ISM)—The dust, molecular clouds, and neutral hydrogen that lie between the stars of this galaxy, generally in the plane of the Milky Way, but whose density is highly variable.

IP/Halley (comet)—The most famous periodic comet. Its aphelion is beyond the orbit of Neptune, but it returns to the inner solar system every 76 years. Named after the 17th-century British scientist, Edmond Halley, who first recognized its regular pattern of reappearances. Studied by a fleet of spacecraft during its 1986 apparition, including the European Space Agency's Giotto.

Isomer—One of two or more substances that have the same chemical composition but differ in structural form.

Kerogens—A family of chemical compounds that make up a portion of the organic matter found in sedimentary rocks. They are insoluble in organic solvents, non-oxidizing acids (HCI and HF), and bases because of their very high molecular weight. Each kerogen molecule is formed by the random combination of numerous monomers. When heated, hydrogen-rich kerogens yield crude oil and hydrogen-poor kerogens yield mainly gas.

Kinetic isotope effect—The effect of a difference in mass between two isotopes of the same element, such as a difference in reaction rate, vapor pressure, or equilibrium constant. The term includes effects on molecular or atomic properties; specific nuclear effects such as radioactivity are excluded.

Kuiper belt—A region of the solar system containing icy planetesimals distributed in a roughly circular disk some 40 to 100 AU from the Sun. Pluto's orbit is believed to circumscribe the innermost region of the Kuiper belt.

Kuiper belt objects (KBOs)—A general name for the bodies found in the Kuiper belt.

Lacustrine—Of or pertaining to lakes.

Murchison (meteorite)—A carbonaceous chondrite, type II (CM2), suspected to be of cometary origin due to its high water content (12 percent).

Neutrino—One of a family of electrically neutral subatomic particles with little or no mass generated during some radioactive decay processes. Because they interact only weakly with matter, neutrinos are extremely difficult to detect.

Organic—Of or relating to any covalently bonded compound containing carbon atoms.

Outgassing—The ejection of gaseous material from the interior of a planet.

Oxidation/reduction—The change in the oxidation state of atoms or ions due to the "loss" or "gain" of electrons.

Photic zone—The richest and most diverse region in Earth's ocean, extending downward to the maximum depth of sunlight penetration, approximately 200 m in the open sea. More generally, the region of a planet's subsurface influenced by solar radiation.

Photolysis—The decomposition of a substance into simpler units as a result of the action of light.

Polar chemical bonds (in organic molecules)—Polarity is induced in chemical bonds when one atom attracts electrons more strongly than the other. A partial separation of charge results. Neither atom bears the full charge of an electron or proton, but one is slightly negative, and the other slightly positive. Such atoms are susceptible to attack in the course of chemical reactions. An atom bearing a partial positive charge is, for example, likely to attract electrons and to participate in the formation of a new chemical bond.

Polycyclic aromatic hydrocarbons (PAHs)—A class of very stable organic molecules made up of only carbon and hydrogen. These molecules are flat, with each carbon having three neighboring atoms, much like graphite. They are a standard product of combustion.

Protein—Any of a group of complex organic compounds, consisting essentially of combinations of amino acids in peptide linkages, that contain carbon, hydrogen, oxygen, nitrogen, and usually sulfur.

Protoplanetary disk—The disk of dust and gas surrounding a star out of which planets form.

Racemic compounds, racemic mixture, racemate—An equimolar mixture of the two enantiomeric isomers of a compound. As a consequence of the equal numbers of levo- and dextro-rotatory molecules present in a racemate, there is no net rotation of the plane of polarized light.

Radiolysis—The breakdown of molecules as a result of exposure to ionizing radiation.

Refractory material—A heat-resistant material that retains its strength at high temperatures, e.g., above the melting point of most metals.

Regolith—The layer of fragmented, incoherent rocky debris on the surface of a planetary body.

Saturated hydrocarbons—Organic molecules containing only single carbon-carbon bonds. As such they cannot incorporate additional atoms into their structure.

SNC meteorites—The family of shergottite, nakhlite, and chassignite stony meteorites believed to have originated on Mars.

Sputtering—A phenomenon occurring when energetic ionized particles impinge on the surface of a solid or liquid target, causing the emission of particles and erosion of the surface of a solid. The sputtered particles from the target can appear as charged or neutral atoms or molecules, atom clusters, or macroscopic chunks of material.

Stereochemistry—The study of how the spatial arrangement of atoms in a compound influences its structural properties.

Stereoisomers—Isomers that differ only in the arrangement of their atoms in space.

Strecker synthesis—The synthesis of α-amino acids by the reaction of an aldehyde or ketone with a mixture of ammonium chloride and sodium cyanide followed by acid hydrolysis of the amino nitriles formed.

Tagish Lake (meteorite)—A unique carbonaceous chondrite collected very soon after falling to Earth in a remote part of northwestern Canada in January 2001.

Tholin—A term used in planetary science to refer generally to organic heteropolymers.

T-Tauri phase—The early phase in the life of a star occurring soon after it has established hydrogen fusion reactions in its core. Such stars are characterized by vigorous surface activity (e.g., flares, eruptions), strong stellar winds, and irregular brightness variations.

Unsaturated hydrocarbons—Organic molecules containing double or triple bonds between adjacent carbon atoms, creating a possibility for further reactions to introduce additional atoms.